Advances in Experimental Medicine and Biology

Volume 1096

More information about this series at http://www.springer.com/series/5584

Heide Schatten

Editor

Molecular & Diagnostic Imaging in Prostate Cancer

Clinical Applications and Treatment Strategies

 Springer

Editor
Heide Schatten
Department of Veterinary Pathobiology
Univ of Missouri-Columbia
Columbia, MO, USA

ISSN 0065-2598   ISSN 2214-8019   (electronic)
Advances in Experimental Medicine and Biology
ISBN 978-3-030-07586-6   ISBN 978-3-319-99286-0   (eBook)
https://doi.org/10.1007/978-3-319-99286-0

Preface

Prostate cancer is still the second most commonly diagnosed cancer among men in the United States, and it is the third leading cause of cancer-related deaths with about 16.5% of deaths resulting from metastatic prostate cancer. Similar statistics have been reported for various countries worldwide.

In recent years, significant progress has been made in developing new imaging modalities for optimal diagnosis and treatment strategies, which have become more personalized and tailored to individual patients. Basic research, improved imaging modalities as well as new clinical trials have opened up new avenues to treat this heterogeneous disease with new possibilities for patient-specific approaches. While significant progress has been made in the early detection of the disease due to improved diagnostic imaging, treatment of advanced stages of prostate cancer is still in the early stages of research but progress is being made due to intense efforts to understand cell migration, epithelial-mesenchymal transition points, and metastasis on genetic, cell, and molecular levels, which has become possible with newly developed research methods, allowing new insights into the disease. Progress has also been made in designing suitable nanoparticles that may be utilized for imaging and targeted prostate cancer treatment. The joint initiatives and efforts of advocate patients, prostate cancer survivors, basic researchers, statisticians, epidemiologists, and clinicians with various and specific expertise have allowed close communication for more specific and targeted treatment. Major forces supporting these efforts are the Department of Defense, the American Cancer Society, and several other foundations that recognized the need for intensified advocacy to find treatments for the disease that represents the most common noncutaneous malignancy for men with new cases resulting in deaths each year.

The present book on *Molecular and Diagnostic Imaging in Prostate Cancer: Clinical Applications and Treatment Strategies* is one of two companion books; the companion book is focused on cell and molecular aspects titled *Cell and Molecular Biology of Prostate Cancer: Updates, Insights and New Frontiers*. The present book includes topics spanning androgen deprivation therapy for prostate cancer; advances in radiotherapy for prostate cancer treatment; role of prostate MRI in the setting of active surveillance for prostate cancer; evaluation of prostate needle biopsies;

multiparametric MRI and MRI/TRUS fusion guided biopsy for the diagnosis of prostate cancer; applications of nanoparticle probes for prostate cancer imaging and therapy; castration-resistant prostate cancer: mechanisms, targets, and treatment; peptide-based radiopharmaceuticals for molecular imaging of prostate cancer; targeted prostate biopsy and MR-guided therapy for prostate cancer; and therapeutic potential of immunotherapy for prostate cancer treatment.

All articles have been selected as invited chapters written by experts in their specific fields who have made significant contributions to prostate cancer research, diagnosis, and treatment, and present the most recent advances in the field. Cutting-edge new information is balanced with background information that is readily understandable to newcomers and experienced scientists and clinicians alike. All articles highlight the new aspects of specific molecular and diagnostic imaging and treatment strategies and on designing new strategies or identifying new targets for therapeutic intervention. The topics addressed are expected to be of interest to scientists, clinicians, students, teachers, and to all who are interested in expanding their knowledge related to prostate cancer for diagnostic, therapeutic, or basic research purposes. The books are intended for a large audience as reference books on the subject.

It has been a privilege and great pleasure to edit this volume titled *Molecular and Diagnostic Imaging in Prostate Cancer: Clinical Applications and Treatment Strategies* and the companion book on cell and molecular aspects, and I would like to sincerely thank all authors and coauthors for their outstanding contributions and for sharing their unique expertise with the prostate cancer community. I hope the chapters will stimulate further interest in finding new diagnostic and treatment possibilities for this disease to increase the health and survival rates of patients particularly of those suffering from metastatic prostate cancer.

Columbia, MO, USA   Heide Schatten

Contents

Contributors

Anyao Bi Xiangya School of Pharmaceutical Sciences, Central South University, Changsha, China

Molecular Imaging Research Center, Central South University, Changsha, China

David Bonekamp Department of Radiology, German Cancer Research Center (dkfz), Heidelberg, Germany

Luís Costa Department of Oncology, Hospital de Santa Maria, Centro Hospitalar Lisboa Norte, Lisbon, Portugal

Oncology Division, Faculdade de Medicina de Lisboa, Instituto de Medicina Molecular, Lisbon, Portugal

Svenja Dieffenbacher Department of Urology, University Hospital Heidelberg, Heidelberg, Germany

Rodney J. Ellis Department of Radiation Oncology, University Hospitals Cleveland Medical Center, Case Western Reserve University, Cleveland, OH, USA

Isabel Fernandes Department of Oncology, Hospital de Santa Maria, Centro Hospitalar Lisboa Norte, Lisbon, Portugal

Oncology Division, Faculdade de Medicina de Lisboa, Instituto de Medicina Molecular, Lisbon, Portugal

Elisha T. Fredman Department of Radiation Oncology, University Hospitals Cleveland Medical Center, Case Western Reserve University, Cleveland, OH, USA

Samuel J. Galgano Department of Radiology, University of Alabama at Birmingham, Birmingham, AL, USA

Tang Gao Xiangya School of Pharmaceutical Sciences, Central South University, Changsha, China

Molecular Imaging Research Center, Central South University, Changsha, China

Giovanna A. Giannico Pathology Medical Director, HCA Midwest Division, Kansas City, MO, USA

Zachary A. Glaser Department of Urology, University of Alabama at Birmingham, Birmingham, AL, USA

Krzysztof R. Gorny Department of Radiology, Mayo Clinic, Rochester, MN, USA

Boris Alexander Hadaschik Department of Urology, University Hospital Essen, Essen, Germany

Omar Hameed Microbiology and Immunology, Vanderbilt University Medical Center, Nashville, TN, USA

Markus Hohenfellner Department of Urology, University Hospital Heidelberg, Heidelberg, Germany

Akira Kawashima Department of Radiology, Mayo Clinic, Scottsdale, AZ, USA

Claudia Kesch Department of Urology, University Hospital Heidelberg, Heidelberg, Germany

The Vancouver Prostate Centre, University of British Columbia, Vancouver, Canada

Xiangqi Kong Xiangya School of Pharmaceutical Sciences, Central South University, Changsha, China

Molecular Imaging Research Center, Central South University, Changsha, China

Michael R. Lewis Department of Veterinary Medicine and Surgery, University of Missouri, Columbia, MO, USA

Yi Liu Xiangya School of Pharmaceutical Sciences, Central South University, Changsha, China

Molecular Imaging Research Center, Central South University, Changsha, China

Daniela Macedo Oncology Department, Hospital de Santa Maria, Centro Hospitalar Lisboa Norte, Lisbon, Portugal

André Mansinho Oncology Department, Hospital de Santa Maria, Centro Hospitalar Lisboa Norte, Lisbon, Portugal

Lance A. Mynderse Department of Urology, Mayo Clinic, Rochester, MN, USA

Vincenzo Pagliarulo Department of Urology, University "Aldo Moro", Bari, Italy

Azienda Ospedaliero-Universitaria Policlinico, Bari, Italy

Tarun K. Podder Department of Radiation Oncology, University Hospitals Cleveland Medical Center, Case Western Reserve University, Cleveland, OH, USA

Kristin K. Porter Department of Radiology, University of Alabama at Birmingham, Birmingham, AL, USA

Jan Philipp Radtke Department of Urology, University Hospital Heidelberg, Heidelberg, Germany

Department of Radiology, German Cancer Research Center (dkfz), Heidelberg, Germany

Soroush Rais-Bahrami Department of Radiology, University of Alabama at Birmingham, Birmingham, AL, USA

Department of Urology, University of Alabama at Birmingham, Birmingham, AL, USA

Heide Schatten Department of Veterinary Pathobiology, University of Missouri, Columbia, MO, USA

Viktoria Schütz Department of Urology, University Hospital Heidelberg, Heidelberg, Germany

Charles J. Smith Department of Radiology, University of Missouri, Columbia, MO, USA

Tamila J. Stott Reynolds Laboratory Animal Resources Center, University of Texas at El Paso, El Paso, TX, USA

David A. Woodrum Department of Radiology, Mayo Clinic, Rochester, MN, USA

Shuiqi Yang Xiangya School of Pharmaceutical Sciences, Central South University, Changsha, China

Molecular Imaging Research Center, Central South University, Changsha, China

Wenbin Zeng Xiangya School of Pharmaceutical Sciences, Central South University, Changsha, China

Molecular Imaging Research Center, Central South University, Changsha, China

Chapter 1
Androgen Deprivation Therapy for Prostate Cancer

Vincenzo Pagliarulo

Abstract In the contemporary scene, less than 5% of men with newly diagnosed prostate cancer (PC) have metastases at first presentation, compared to 20–25%, more than 20 years ago. Nonetheless, the use of androgen deprivation therapy (ADT) has increased over the years, suggesting that patients in Europe and United States may receive ADT in cases of lower disease burden, and not always according to evidence based indications. Nonetheless, PC remains the second most common cause of cancer death after lung cancer in American men. Thus, there is a need for more effective, specific and well tolerated agents which can provide a longer and good quality of life while avoiding the side effects related to disease and treatment morbidity.

After mentioning the current knowledge on the endocrinology of androgens and androgen receptor, relevant to PC development, as well as the possible events occurring during PC initiation, we will compare different hormonal compounds available for the treatment of PC, both from a pharmacological standpoint, and in terms of contemporary clinical indications.

1.1 Historical Perspective

Starting 1941, patients with metastatic prostate carcinoma have been receiving hormonal manipulation to suppress the endogenous production of androgens [1]. Androgen deprivation therapy became widely accepted as the treatment of choice for the palliation of advanced prostate cancer (PC), as it provided dramatic relief of pain and symptoms, reduction of associated risks (bone fracture, urinary retention), and delay of disease progression to a symptomatic stage. For several decades, ADT was achieved by surgical castration or suppression of gonadotropin releasing

V. Pagliarulo (✉)
Department of Urology, University "Aldo Moro", Bari, Italy

Azienda Ospedaliero-Universitaria Policlinico, Bari, Italy
e-mail: vincenzo.pagliarulo@uniba.it

© Springer Nature Switzerland AG 2018
H. Schatten (ed.), *Molecular & Diagnostic Imaging in Prostate Cancer*,
Advances in Experimental Medicine and Biology 1096,
https://doi.org/10.1007/978-3-319-99286-0_1

hormone (GnRH) production at the level of the hypothalamus with diethylstilbestrol (DES) [2]. Only in 1959 the Veterans Administration Cooperative Urological Research Group [3] began their investigations into the effects of DES, revealing that a daily administration of 5 mg DES was associated with prohibitive cardiovascular toxicity; importantly, 1 mg DES was equally effective to 5 mg without the cardiovascular effects, however, castrate levels of testosterone were not reached reliably [4]. As the primary objectives of ADT were to avoid surgical castration, achieve castrate levels of testosterone, and minimize cardiovascular toxicity, a daily dose of 3 mg DES was thought to best achieve these objectives and ultimately became the accepted regimen for pharmacologic castration [5].

During the 1970s, Schally purified the GnRH decapeptide, started the synthesis of its agonists, and was finally awarded with the Nobel Prize for discovering the basis of medical castration [6]. The first randomized clinical trial comparing leuprolide to 3 mg of DES in patients with metastatic PCa, would show equivalence in reducing serum testosterone to castrate levels [7]. However, leuprolide caused fewer thromboembolic and other side effects than DES; ultimately, synthetic GnRH agonists have replaced DES and orchiectomy as the preferred approach to androgen deprivation. In the effort to decrease the side effects of ADT other treatment modalities have been developed over the years. Nonsteroidal antiandrogens (bicalutamide, flutamide, nilutamide) competitively inhibit the binding of androgens to the androgen receptor (AR), without affecting serum testosterone levels, and finally showing a more favorable toxicity profile compared to castration. In 2003, abarelix was the first GnRH antagonists to receive approval for advanced, symptomatic PC. However, abarelix was promptly removed from the market because associated to the risk of developing life-threatening, systemic allergic reactions. After extensive clinical testing in patients with advanced PC, degarelix was shown to be safe and to retain the same therapeutic properties observed with previous LGNRH antagonists, thus receiving FDA approval in 2008 [8]. Importantly, compared to LGNRH agonists, degarelix was able to avoid testosterone surge and achieve castrate serum testosterone levels within few days [9]. In patients with advanced and metastatic PC, ADT is more commonly referred to as a palliative therapy, being unable to warrant cure from cancer. Several mechanisms are triggered by PC in order to activate cell proliferation, disease progression, and metastatic spread, regardless of androgen suppression obtained with the above-mentioned drugs. Initially, this condition was called "androgen insensitivity", as it was thought tumor cells would escape the need of androgens for their growth. In more recent years, this term has been replaces with "castration resistance", as a new wave of research has demonstrated that PC continues to be hormone-dependent even when evolving to castration resistance. In fact, continued androgen synthesis following ADT occurs in the testes, adrenals and the tumor itself, thru activation of alternative pathways. These findings have led to the discovery of newer targets for androgen suppression, and finally, the development of second line androgen deprivation therapies that have successfully prolonged the life of patients with advanced PC.

1.2 Sex Steroids in Prostate Cancer

1.2.1 Sources of Androgens

Testosterone synthesis is controlled by the pulsatile release of a hypothalamic deca-peptide called gonadotropin-releasing hormone (GnRH), which, in turn, acts at the pituitary gland stimulating the release of luteinizing hormone (LH) and follicle-stimulating hormone (FSH) into the peripheral circulation. In the testes, FSH binds to Sertoli cells and is required for induction and maintenance of spermatogenesis, while LH binds to Leydig cells and stimulates the secretion of gonadal sex steroids. Leydig cell secretion creates a high local concentration of testosterone in the testis, however, testosterone is also secreted into the circulation, being the major feedback signal that controls the physiological operation of the hypothalamic-pituitary axis. Specifically, when testosterone levels are sufficient, the pituitary gland decreases the production and release of LH, which also inhibits hypothalamic GnRH secretion.

The testes are not the only source for androgens; although they contributes to the production of more than 95% of total circulating testosterone in the adult male, the adrenal glands are responsible for the release of other circulating sex steroids. Dehydroepiandrosterone (DHEA) and androstenedione are produced in the zona reticulata and zona fasciculata of the adrenal cortex and secreted in large amounts in the bloodstream, as circulating levels of DHEA are more than 100 times higher than testosterone [10]. They are released as inactive precursor steroids and distrib-uted indiscriminately to many peripheral tissues; the transformation of DHEA into androgens, however, is tissue-specific, reaching high levels in the prostate, where DHEA is converted in testosterone and dihydrotestosterone (DHT). Androgens of adrenal origin are critical and may drive castration resistant PC as it was shown that in castrated men an important proportion of androgens is still present in the human prostate which is made from the DHEA present in the prostate itself [11]. Importantly, testosterone and DHT produced after DHEA conversion exert a strictly local action without significant release of active sex steroids in the circula-tion, thus without affecting serum levels of testosterone. The contribution of adre-nal DHEA to total androgen levels is best illustrated by the concentration of intraprostatic DHT and androstenedione remaining in the prostate after castration [12]. These data show that of the total concentration of DHT measured in the pros-tate of non-castrated men, up to 50% is still present after castration. Further, no decrease in prostatic levels of androstenedione was found after castration [13], which is of particular significance as this androgen has been shown to bind wild type AR without being inhibited by flutamide or bicalutamide [14]. These findings clearly demonstrate that achieving castrate levels of circulating T does not elimi-nate androgens from the prostate tumor microenvironment. Residual tissue andro-gens are implicated in driving the majority of mechanisms whereby persistent AR-mediated signaling drives castration resistant disease. These data are relevant to explain the exciting development of new generation hormonal compounds, and most importantly, the significant results obtained in the last decade after their introduction into clinical practice.

1.2.2   Androgens and the Androgen Receptor

Once released in the bloodstream, testosterone and other less active sex steroids reach their peripheral sites of action, where they undergo reversible and irreversible metabolism to other steroids with different activities. Specifically, testosterone is converted to DHT by the action of 5α reductase in target tissues; although it is about one tenth as abundant as testosterone, it accounts for most of testosterone's biological action. Male sex hormones have a wide range of functions, including promoting the development of male primary and secondary sexual characteristics, stimulating erythropoiesis, increasing metabolic rate, increasing bone density and stimulating libido [15].

The normal physiologic function of androgens is a result of stimulating the AR. The AR is a member of the nuclear hormone receptor family of transcription factors, which also includes the estrogen, glucocorticoid, progesterone and others receptors comprising of four distinct functional domains [15]. In the absence of a ligand, AR resides primarily in the cytoplasm in association with heat shock proteins (HSPs), cytoskeletal proteins and other chaperones. Binding of a ligand to the AR induces conformational changes in the ligand binding domain and a well-described series of events finally resulting in traslocation to the nucleus and binding to specific recognition sequences known as "androgen response elements" (AREs) within the promoter region of AR target genes thereby modulating gene expression.

The AR is expressed in many tissues including the prostate and testes [17]. In the *testes*, androgens play a critical role during puberty. During this phase gonadotropins are responsible primarily for the formation of the adult cohort of Sertoli, Leydig and stem germ cells and their functions that will eventually lead to normal spermatogenesis and sperm production. Thus, hormone deprivation during puberty will affect the normal scrotal descent and development of the adult testis. In contrast, in the adult, the effects of hormone deprivation are essentially on the germ cells composition via functional impairments in the somatic cells, particularly the Sertoli cells.

Within the *normal prostate*, AR plays a key role in stromal and epithelial cells. The stroma surrounds the glands within the prostate and is responsible for producing many of the factors that regulate the growth and development of prostatic epithelial cells [18, 19]. Prostatic epithelial cells shape glands within the prostate, which contain luminal/secretory, basal, and neuroendocrine cells [20]. Luminal cells express high levels of AR and respond directly to androgens by stimulating production and secretion of prostatic differentiation markers. Basal cells are usually considered to lack AR expression. Several lines of evidence suggest that in the normal adult prostate stromal AR signaling promotes epithelial cell growth and prostate regeneration. Once activated, AR signaling functions to modulate the secretion of different growth factors, including vascular endothelial growth factor (VEGF), fibroblast growth factors (FGFs) [21, 22], and survival factors, which diffuse through the stroma and act on epithelial cells to facilitate prostate growth and cell death [23]. On the other side, epithelial AR signaling functions to maintain homeostasis through suppression of basal cell proliferation and stimulation of differentiated luminal cell survival [23, 24]. These studies establish the homeostatic regulation of the adult prostate gland, and provide the basis against which to measure changes that occur in the androgen/AR signaling axis during the development of PC.

Finally, androgens and AR activation play a critical role in *prostate cancer*, although most of the evidence derives from clinical contributions, while direct evidence for the necessity of AR in prostate tumorigenesis is scarce. One possible key event in PC initiation is a gain of function of luminal prostatic cells in which the AR is able to engage the molecular signaling pathways driving proliferation and survival of these cells. In fact, PC tissue is primarily composed of luminal epithelial cells in which AR plays a suppressive role during adult prostate homeostasis, however this role changes after malignant transformation. The events involved in this 'malignancy switch' are poorly understood, however several changes that occur in AR signaling have been reported and reviewed [25]. Androgens play a crucial role in PC development as well. Both preclinical and clinical studies have demonstrated a correlation between serum testosterone levels and the risk of developing PC, being higher levels protective in reducing this risk [26, 27]. One hypothesis is that low systemic levels of testosterone may cause AR within prostate luminal cells to regulate expression of genes that turn cells to be less dependent on growth and survival. Further, low testosterone levels may induce mutagenesis within prostatic stem cells, thus favoring tumorigenesis.

The important role of androgens is also evident from PC preventive trials in which the prolonged use of a 5a-reductase inhibitor (finasteride and dutasteride) caused a 23–24% risk reduction of developing cancer [28, 29]. Importantly, these drugs reduce the tissue but not the systemic levels of testosterone. These findings clearly demonstrate that androgens within the prostate microenvironment may play a central role. Residual tissue androgens may be implicated in driving the majority of mechanisms leading to PC initiation and justify the persistent AR-mediated signaling that leads to castration resistant disease. From a molecular standpoint, few studies have focused on the biological effects of androgens withdrawal on PC cells. Within the normal prostate, basal cells are androgen responsive but are not dependent on androgen for survival, whereas secretory cells require androgen to avoid apoptotic cell death. Similar to normal prostate, PC cells require androgens for continued growth. Although the mechanisms for the clinical response to ADT are not clearly defined, androgen deprivation leads to apoptosis of the secretory epithelium and growth arrest of the basal epithelium in the normal prostate [30]. These events are coupled to marked changes in several of the following parameters: tumor cell nuclear area, cell proliferation rate, apoptosis rates, and vacuolization of the tumor cell cytoplasm [31, 32].

1.3 Pharmacology of Hormonal Therapy for Prostate Cancer

In patients with PC requiring first line medical intervention, the primary biochemical goal is to effectively suppress the transcriptional potential of AR, thru a range of treatment options collectively referred to as hormonal therapy or as androgen deprivation therapy (ADT). Depletion of androgens can be achieved either by suppressing the secretion of gonadal androgens, by means of bilateral orchiectomy or pharmacological castration, or by administering "antiandrogens" that inhibit the action of circulating androgens at the level of their receptor. However, achieving castrate levels of circulating T does not eliminate androgens from the prostate tumor microenvironment. As previously discussed, residual tissue androgens are implicated in the majority of mechanisms driving PC progression. This has been the rationale for using 5α reductase inhibitors to block testosterone conversion to DHT within prostatic tissue, and more importantly, to explain the exciting development of the new hormonal compounds, such as abiraterone end enzalutamide. The optimum serum castration levels to be achieved with ADT are still debated. Although there are recognized limitations in measuring serum testosterone concentrations [33], a total testosterone concentration >300 ng/dl (10.4 nmol/l) is generally considered normal [34]. The upper limit of castration concentrations of serum testosterone is considered to be 50 ng/dl (1.7 nmol/l), although lower concentrations (20 ng/dl; 0.7 nmol/l) may be more desirable for optimal therapy [35].

1.3.1 Gonadal Androgen Ablation

Bilateral orchiectomy. Bilateral orchiectomy is a relatively simple procedure that causes the ablation of testicular androgens and has been a reference for comparison of medical ADTs, both in terms of the circulating testosterone to be reached (≤20 ng/dl) [36] and the castration pattern, which is rapid and sustained after orchiectomy. Although free of compliance issues and apparently associated with good quality of life (QoL) [37], bilateral orchiectomy has fallen out of favor and is now largely replaced by medical castration. Furthermore, a systematic review and meta-analysis has shown that survival after therapy with a GnRH agonist is equivalent to that after orchiectomy [38].

    GnRH agonists. A GnRH agonist is a synthetic peptide modeled after the hypothalamic neuro-hormone GnRH that interacts with the GnRH receptor to elicit its biologic response. Compared to the wild type GnRH, agonists are modeled with specific aminoacid substitutions typically in positions 6 and 10. Several GnRH agonists are currently available, as well as different extended-release formulations of these drugs. GnRH agonists with two substitutions include leuprolide, buserelin, and goserelin. Triptorelin is an agonist with only a single substitution at position 6. Medical castration with GnRH agonists was an important discovery in medicine, as an inhibitory effect on testicular production of testosterone was observed unexpectedly using a stimulatory molecule. In fact, the mechanism of action of these compounds is based on the overproduction of LH following their binding to the GnRH receptors in the pituitary cells. This results in a surge of circulating testosterone for a short period of time (days or weeks), known as the "flare reaction" [39]. In patients with advanced disease, this transient testosterone surge may translate into an increase of symptoms, known as disease flare, such as bone pain at the site of metastasis, urinary tract obstruction, or severe neurologic complications of spinal metastasis. This phenomenon may be ameliorated by prescribing nonsteroidal antiandrogens, such as flutamide or bicalutamide, for the first 2–3 weeks, prior or during the first administration of GnRH agonists [40]. After transient elevation in circulating testosterone, chronic exposure to GnRH agonists will result in a downregulation of GnRH receptors and a downward suppression of testosterone production will follow. Administration of a GnRH agonist will cause testosterone to reach castration levels generally within 2 weeks; however, the duration of testosterone suppression after a single injection is variable among patients, depending mainly on the formulation used and on individual variability. In many studies, authors have selected arbitrarily injection intervals of 3–4 months using goserelin, buserelin or leuprolide, however castration levels are maintained for a considerably longer time than the nominal interval for treatment repeated dose. Time for testosterone recovery after medical castration will be discussed later on. Ideally, the goal of androgen ablation is to consistently achieve and maintain the lowest testosterone levels possible without the unpredictable rises in testosterone. There have been several reports as well as systematic reviews on testosterone levels reached during medical castration with GnRH agonists in patients with PC [41]. According to these data, when a cutoff point of 50 ng/dl is defined as a castration level of testosterone, GnRH agonists allow castration in most of the patients, whereas, with regards to the 20 ng/dl breakpoint, great differences achieving castration appear among the reports. In an observational study, serum testosterone was monitored every 6 months in 73 patients starting therapy with three monthly depot formulations of GnRH agonists between 2001 and 2003 [42]. At the first determination, 39.7% failed to achieve serum testosterone <20 ng/dl, and 12.3% failed to reach <50 ng/dl. These percentages rose at the second determination (50.7% and 19.2%, respectively) and tended to stabilize thereafter. Importantly, the first three determinations could be used to predict subsequent testosterone elevation. Finally, GnRH agonist formulations may differ in testosterone suppression levels and duration of suppression; however, the clinical efficacy of these compounds is hard to compare, as no data relate these differences to differences in disease progression and survival. In fact, no study has prospectively evaluated the clinical implications of an incomplete castration; nonetheless, some assumptions could be made starting from intermittent androgen therapy trials.

GnRH antagonists. Testosterone suppression can also be achieved with the use of a GnRH receptor blockers and the clinical efficacy of these agents is now well established. These drugs, commonly called GnRH antagonists, represent a new class of hormonal therapy that induce a faster suppression of serum testosterone than GnRH receptor agonists, but without a testosterone surge. The principal mechanism of action of GnRH antagonists is competitive receptor occupancy of GnRH-r. After 30 years of trial and error, degarelix was approved by the USA FDA, and by the EMEA for the treatment of advanced PC. Degarelix is administered at an initial dose of 240 mg followed by a monthly maintenance dose of 80 mg, as these were shown to be the best to achieve a rapid, profound, sustained suppression of testosterone in more than 90% of patients within 3 days and in more than 95% of patients 1 month after injection [43]. In a recent phase III trial (CS21), both tested dose regimens of degarelix (240/80 and 240/160 mg) and leuprolide 7.5 mg suppressed testosterone to ≤0.5 ng/ml in >95% of patients over a 1-year treatment period [9]. Both degarelix regimens achieved a more rapid reduction of testosterone and PSA than leuprolide, and neither degarelix dose induced testosterone surge or microsurges. The most commonly observed side effects related to degarelix are injection-site reactions (pain, erythema, swelling). Otherwise, degarelix is generally well tolerated, without systemic allergic reactions, frequently observed with previous GnRH antagonists, and with most adverse events consistent with the underlying condition.

1.3.2 Androgen Receptor Antagonists

Shortly after the initial discovery of the AR in the late 1960s, these drugs have historically been included in the backbone of PC therapy. For a long time, first generation AR antagonists have represented the best available hormonal therapy for advanced prostate tumors. Subsequently, their role has been downsized by the introduction of GnRH agonists. Steroidal (cyproterone acetate, megestrol acetate) and non steroidal (flutamide, bicalutamide and nilutamide) AR antagonists or antiandrogens (AAs) serve as oral competitive inhibitors to sex steroids, the endogenous ligands to the AR.

Cyproterone acetate is a synthetic derivative of 17-hydroxyprogesterone, which functions as an AA, however it may also reduce serum testosterone, as it inhibits androgen production in the adrenals due to its progestational activity. Thus, the use of cyproterone aceteate, in contrast to non-steroidal antiandrogens, may result in the suppression of libido and erectile function. For the treatment of advanced PC, cyproterone acetate may be administered prior or in combination to GnRH agonists at a maximal oral dosage of 300 mg/day. In patients under ADT experiencing hot flushes, it is often administered to reduce this side effect, again thanks to its progesteronic properties, at a maximal dosage of 150 mg/day.

Non steroidal AA do not reduce circulating levels of testosterone, on the contrary these may increase as LH and FSH release may be excited to compensate for androgen blockade. There is a lack of robust data regarding the use of flutamide and nilutamide in patients with advanced and metastatic PC, as most studies have included a limited number of patients. Nowadays, bicalutamide is the most commonly used antiandrogen, in combination with GnRH agonists and, to a lesser extent, in monotherapy, because of its long half-life (about 6 days), higher potency and better tolerability when compared to flutamide and nilutamide [44]. It is usually administrated at the dose of 50 mg/day when used as part of the CAB, and at the dose of 150 mg/day when used as single agent.

1.3.3 Abiraterone

As previously mentioned within this chapter, the testes are not the only surce of androgens. Dehydroepiandrosterone (DHEA) and androstenedione are produced in the zona reticulata and zona fasciculata of the adrenal cortex and secreted in the bloodstream. Once target organs are reached, such as PC tissue, DHEA is converted in T and DHT.

Abiraterone acetate is the oral prodrug of abiraterone, a specific inhibitor of steroidogenesis. Specifically, abiraterone inhibits cytochrome P450 c17, a rate-limiting enzyme in androgen biosynthesis, which has two distinct activities: lyase and alpha hydroxylase. This enzymatic action is used in the conversion of pregnenolone and progesterone to 17-OH pregnenolone and 17-OH progesterone, and from there to dehydroepiandrosterone and androstenedione, which is the penultimate step in testosterone production. Thus, this drug interrupts androgen production at three sources: the testis, the adrenal glands and the tumor itself [45]. Use of abiraterone to inhibit androgen synthesis, however, is associated with several undesired physiologic changes, including a decrease in cortisol levels and a compensatory increase in adrenocorticotropic hormone (ACTH) [46]. This rise in ACTH leads to accumulation of steroids with mineralocorticoid properties upstream of cytochrome P450 c17 in the cortisol biosynthetic pathway and, ultimately, to mineralocorticoid-related adverse events, including hypertension, hypokalemia, and fluid retention. When coadministered with abiraterone acetate, low-dose prednisone or prednisolone substitutes for cortisol, compensating for the abiraterone-induced reduction in serum cortisol. Currently, abiraterone acetate is approved only for use in combination with the prednisone or prednisolone dose given orally. Given the long-term administration of prednisone in combination with abiraterone acetate, there is the potential for high levels of total corticosteroid exposure, placing patients at risk of corticosteroid-related adverse events (most frequently: edema, hypertension, weight gain, hyperglycemia and steroid-induced diabetes).

1.3.4 Enzalutamide

The new generation antiandrogen Enzalutamide, previously known as MDV3100, was selected from a library of compounds under clinical development, because of its favorable drug-like properties, its effect on castration-resistant PC (CRPC) xenograft models, and, most importantly, because of its ability to inhibit AR signaling as a result of high binding affinity to the AR and lack of agonist activity. In fact, enzalutamide binds the AR with an eightfold greater affinity compared to bicalutamide, when evaluated using an 18-fluoro-deoxyglucose-dihydrotestosterone scan to measure relative AR binding affinity in a competition assay [47]. In addition to an increased binding affinity, enzalutamide inhibits the AR translocation into the nucleus and AR-mediated transcription and cell growth in vitro, while bicalutamide does not. Further, Enzalutamide induces regression of established LNCaP/AR xenograft tumor cells growing in castrated male mice, while bicalutamide treatment only slows tumor growth. Regression seen with enzalutamide is associated with continued evidence of apoptosis up to 25 days after initiation of treatment [47]. Finally, gene expression profiling in LNCaP cells indicated that enzalutamide opposes agonist-induced changes in genes involved in processes such as cell adhesion, angiogenesis, and apoptosis [48]. These data indicate that MDV3100 may be a true AR antagonist without partial agonist properties. Administered at a daily dose of 160 mg, enzalutamide seems to be very well tolerated with a favorable side effect profile. Expected toxicities, as resulted from the AFFIRM study [49], include fatigue, diarrhea, musculoskeletal pain, headache, hypertension, and hot flashes.

1.3.5 Five Alfa Reductases Inhibitors

Dihydrotestosterone (DHT) is an endogenous androgen sex steroid and hormone. The enzyme 5α-reductase catalyzes the reduction of testosterone into DHT in several tissues, including the prostate gland. Importantly, DHT is the most potent known endogenous ligand of the AR, as it's affinity for the human AR is about two- to threefold higher than that of testosterone [50] and 15–30 times higher than that of adrenal androgens. In addition, the dissociation rate of DHT from the AR is fivefold slower than that of testosterone [51]. Five α-reductase inhibitors (5ARI) like finasteride and dutasteride inhibit 5α-reductase type II and/or other isoforms causing a decrease in circulating DHT levels. However, DHT plays only a minor role as a circulating hormone as it acts mainly in an intracrine and paracrine manner in the prostate in which it is produced. Interestingly, one study showed that treatment with dutasteride resulted in almost complete suppression of intraprostatic DHT, increased apoptosis, and decreased microvessel density [28]. Although 5α-Reductase inhibitors were developed and are used primarily for the treatment of BPH, as these drugs are able to significantly reduce the size of the prostate gland and alleviate BPH symptoms, these findings suggest that 5ARI can cause regression in PCa. Another study reported that after short-term dutasteride treatment, benign epithelium showed involution and epithelial shrinkage, and PC tissue demonstrated a decrease in epithelium relative to stroma [52]. These findings indicate that dutasteride induces significant phenotypic alterations in both benign and neoplastic prostate, supportive of a preventive or therapeutic role.

1.3.6   ADT and FSH Escape

Several studies have demonstrated that when GnRH agonists are administered, FSH production is suppressed to a lesser extent as compared to LH. Further, after few weeks, FSH levels start rising until reaching baseline circulating concentrations. This event is called "FSH escape" and is also observed after bilateral orchiectomy and treatment with AR antagonists. In contrast, the drop in serum FSH in patients receiving GnRH antagonists was found to be persistent after 1 year of treatment [9]. The clinical relevance of this phenomenon is unclear, however FSH receptor (FSH-R) may be involved in the response to LGNRH agonist and antagonist activity. Several expression studies have found that FSH-R is more abundant in prostate adenocarcinoma clinical samples compared to normal prostate tissue and BPH glands; within PC cell lines, androgen-insensitive cells express FSH-R, while sensitive cells do not [53]. These reports may suggest that FSH and its receptor could play an important role in the progression of PC as an autocrine or paracrine factor and that FSH-R stimulation could have a proliferative function in the castrate-refractory tumour state [54]. Whether GnRH antagonists are more effective compared to GnRH agonists in reaching this goal, as a result of a better FSH activity suppression, needs further investigation in the clinical setting.

1.3.7   Testosterone Recovery After Castration

While medical castration is generally believed to be reversible, there are only a few published trials that specifically study the kinetics of the increase of testosterone. The kinetics of the normalization of androgens after ADT has significant implications for designing and interpreting treatment schemas that include discontinuing GnRH agonists such as neoadjuvant, adjuvant and intermittent therapy. Androgen recovery is dependent to age and duration of castration. Younger age is associated with a faster recovery of testosterone to supracastrate, and baseline or normal levels [55, 56]. One plausible explanation for an inverse relationship between age and testosterone recovery is a diminishing function of the hypothalamic-pituitary axis and/or Leydig cells with age. Testosterone levels decrease as men age and an average rate of decrement of 3.7 nmol/l per decade was estimated [57]. Thus, the decreased ability of older men to produce and secrete testosterone is a possible contributing factor for the decreased testosterone recovery in older patients. Further, the time course of testosterone recovery is generally related to the duration of androgen withdrawal [56, 58, 59]. The longer the duration the longer it takes for testosterone levels to recover. In a large study of patients undergoing only 6 months of GnRH agonist therapy for biochemical recurrence DHT and T levels did not return to normal for another 14.9 and 16.6 weeks, respectively [60]. In a prospective study on buserelin 9.45 mg, 3 months implant, drug release was maintained for 230 days and time to return to 50% of baseline testosterone levels was about 10 months [61]. In a large study evaluating 267 men undergoing ADT via various methods for a median of 9 months, at 3 years 79% of patients had testosterone levels recovered to normal [62]. Although older age and a low pretreatment testosterone level were associated on multivariate analysis with a delayed recovery to normal testosterone levels, duration of castration was not. Regardless of what are the causes or predisposing factors, the prolonged testosterone recovery after androgen withdrawal has several clinical implications. In clinical studies involving ADT, knowledge of testosterone recovery must be taken into consideration for the interpretation of PSA relapse-free rates as PSA is an androgen dependent marker. A premature analysis before an adequate testosterone recovery will lead to an overestimation of PSA relapse-free rates. Likewise, when short courses of ADT are used, as in patients undergoing radiation therapy, the use of immediate post-treatment PSA values for prognosis may not be appropriate since testosterone and PSA levels may be suppressed for a while after treatment.

1.4 Hormonal Therapy in Non-metastatic Prostate Cancer

In the setting of patients diagnosed with locally or locally advanced PC, without any clinical or radiographic evidence of metastases, the use of hormonal therapy is more controversial as compared to patients with more advanced and metastatic PC. Aside from the proven role in combination to radiation therapy in intermediate and high risk patients, in other clinical scenarios, the evidence in favor of adopting androgen deprivation is less robust. Nonetheless, an increasing number of clinicians and patients have turned to androgen deprivation therapy as an alternative to surgery, radiation, or conservative management, especially among older men.

1.4.1 Hormonal Therapy Alone

The use of ADT as sole PC therapy is also called primary ADT (PADT). Despite uncertainty about its impact and potential toxicity, PADT has been applied to a significant proportion of patients with organ confined and locally advanced PC [63, 64]. Population based studies and randomized trials have been carried out to understand which patients may benefit from primary ADT. Interestingly, studies comparing primary ADT to observation, both among patients with clinically localized [65–67] and clinically advanced disease [68], have shown no survival advantage in patients receiving ADT. The Early PC (EPC) programme evaluated the addition of daily bicalutamide 150 mg to standard of care in the M0 disease. Among patients with both localized and locally advanced PC, primary bicalutamide did not improve overall survival (OS) compared to observation [69].

Similarly, three trials have suggest that adding radiation therapy of the primary tumor to ADT improves survival compared to ADT alone in patients with high risk PC [70–72]. Finally, in patients with clinical evidence of N+ disease, radical prostatectomy seems to be superior to ADT alone [73–75]. In conclusion, in patients with M0 disease, hormonal therapy alone is not favorable, as it is comparable to observation, and inferior to other primary treatments. In this setting, PADT may be recommended only for men unfit for any other standard treatment, if they have symptoms related to locally advanced disease.

1.4.2   Hormonal Therapy in Combination to Radiotherapy

The use of ADT in combination to external beam radiation therapy adds overall survival benefits compared with RT alone. A first hypothesis is that this may derive from a better local control of PC. From an experimental standpoint, preclinical studies give a strong support to the neoadjuvant combination of AD and RT [76]. It was shown in murine models that orchiectomy performed prior to RT, rather than afterwords, produces a significantly greater decline in the total dose required for 50% tumor control. Possible biological explanations may be related to the role of hypoxia in decreasing the susceptibility of tumor cells to the damage induced by radiation. Milosevic et al. investigated the impact of ADT on hypoxia in human prostate tumors and were the first to demonstrate that androgen deprivation increases the oxygen levels in prostatic tumoral tissue [77]. Afterwords, it was shown that while androgens increase the basal levels of superoxide, androgen depletion inhibits the NADPH oxidase system and acts as a radiosensitizer, enhancing the vulnerability of PC cells to toxic oxidative stress induced by radiations [78].

Not all patients undergoing radiotherapy will benefit from the addition of ADT. Findings from observational data and subgroup analyses within several randomized trials confirm that this approach does not turn in any benefit for lower-risk patients (cT1–2a, Gleason score 2–6, PSA <10 ng/ml) [79–83]. Among patients with intermediate risk disease (cT2b, Gleason score 7, PSA 10.1–20 ng/ml) the combination of conventional dose RT (<72 Gy) with short-term ADT (4–6 months) has been reported to improve local control and survival [84–87]. Data from RTOG 08–15 are awaited and will provide insights to the role of combining ADT with dose escalated RT. Patients undergoing RT at high risk or with nodal disease (cT2c, Gleason score 8–10, PSA >20 ng/ml) are those who benefit the most from an immediate and long term ADT. Overall, several randomized controlled trials have proven a significant survival advantage among patients receiving long term (2–3 years) ADT plus RT, either compared to those patients receiving RT alone, and to those receiving shorter exposure to ADT [88, 89]. The majority of studies mentioned so far have used GnRH agonists, with or without an antiandrogen, for the combination of hormonal treatment to RT; there is minor clinical evidence regarding the use of degarelix or of an AR antagonist.

1.4.3 Hormonal Therapy Adjuvant to Radical Prostatectomy

The role of immediate androgen deprivation for patients at high risk for recurrence after local treatment as opposed to a delayed treatment is an old controversy that remains unresolved at the present time. Over the last decades, we have learned that many men who are not cured by radical prostatectomy have no symptoms and may have long life expectation, even when PSA starts to increase [90, 91]. One study that has largely influenced clinical practice is the Eastern Cooperative Oncology Group (ECOG) 3886 trial [92]. Patients who underwent RP and had pathological evidence of nodal involvement were randomized to either immediate post-operative or deferred (at the time of bone metastases) ADT. Although a significant improvement in OS (p = 0.04) and CSS (p < 0.0001) in favor of the patients receiving ADT immediately after surgery was found, the study raised several concerns [93], such as small sample size (98 men from 36 centers), the lack of a central pathologic review to assess stage and Gleason scores, and the lack of baseline PSA testing. More recently, adjuvant endocrine therapy after surgery was evaluated in the EPC trial. Interestingly, in a retrospective cohort study based on the USA-Medicare data, patients who underwent radical prostatectomy and had positive regional lymph nodes were analyzed. When men receiving immediate adjuvant ADT were compared to those not receiving adjuvant ADT no statistically significant survival difference were seen, using propensity scores to balance potential confounders. As the authors speculate, routine post-RP monitoring of PSA and biochemical recurrence (BCR) as the trigger for ADT initiation in the delayed arm, may partly justify the different results compared to ECOG 3886 [66, 67].

1.4.4 Hormonal Therapy at Biochemical Relapse After a Primary Treatment

Among patients undergoing surgery with a curative intent, nearly one third will fail and face a situation of PSA only recurrence [90]. Among these patients survival can be long [91], nonetheless, detection of BCR is a contemporary trigger to start ADT, despite the fact that there are no RCTs demonstrating a benefit in survival or quality of life. Two studies have retrospectively compared clinical outcomes in patients receiving or not ADT at the time of biochemical recurrence after radical prostatectomy [94, 95]. In both cases, an advantage towards the cohort receiving ADT was seen only in higher risk patients, and only relative to cancer specific survival. The fact that an advantage in cancer specific survival does not bring to an OS benefit, might introduce the issue of toxicity related to lifelong ADT. Therefore, ADT in patients experiencing BCR after surgery should not be administered outside clinical trials.

A treatment option with limited side effects in men with biochemical progression after radical therapy could be the use of 5α-reductase inhibitors (5-ARI).

Three randomized clinical trials have investigated the role of both finasteride and dutasteride in this setting [96–98], showing that 5-ARI are able to delay further PSA progression and clinical-related outcomes, nonetheless, survival endpoints were not taken into account. Finally, additional data would be helpful in deciding if 5-ARI could be considered a treatment option in this setting.

1.5 Hormonal Therapy in Metastatic Prostate Cancer

Depending on their risk profile, a proportion of patients with localized or locally advanced PC will eventually experience distant recurrence, while approximately 5% of patients are metastatic at the time of PC diagnosis. Nonetheless, probably as a result of improved ability to diagnose metastases at an earlier stage, the incidence of metastatic PC is increasing, mainly among younger men [99]. There is little debate regarding the immediate need for hormonal therapy in these patients. In this setting, the risk of developing symptoms (bone pain, renal failure, anemia, pathologic fractures, spinal cord compression) can be reduced with early implementation of ADT. Over the years, many debates have surrounded ADT: early versus deferred initiation, the addition of an antiandrogen for combined androgen blockade (CAB) versus monotherapy, GnRH agonists versus antagonists, and intermittent therapy versus continuous therapy. Regardless, it is known that 80–90% of patients will initially respond both clinically and biochemically to ADT, and this translates into disease control for several years and improvements in cancer-related symptoms. Nonetheless, hormonal therapy is rarely curative, and in the metastatic setting, cancer typically progresses within 2–3 years despite castrate levels of serum testosterone [100]. This stage is known as castration-resistant PC (CRPC) and, despite many more treatment options, commonly leads to death in 2–4 years [101].

Recent breakthroughs in the understanding of the mechanisms of PCa adaptation to ADT, focusing on the AR pathway, have demonstrated that PCa continues to be hormone-dependent even when evolving to castration resistance. In CRPC, continued androgen synthesis following ADT occurs in the testes, adrenals and the tumor itself, with consequent ongoing activation of AR signaling [102, 103]. The development of novel agents targeting the androgen axis with different mechanisms of action offered a new therapeutic option to further suppress androgen levels and prolong life in patients with CRPC. Among others that still are under clinical investigation, two drugs are now available that have demonstrated survival advantage in patients with CRPC, both in the post and pre chemotherapy setting [49, 104–106]. These agents are enzalutamide and abiraterone acetate plus prednisone, and their use is now strongly supported by all society guidelines in the setting of CRPC. Importantly, these agents are being now tested in randomized phase III clinical trials in the setting of hormone sensitive metastatic and non metastatic PC. Ground-breaking results have already been published showing survival advantages for the use of abiraterone acetate plus prednisone over ADT alone in this category of patients and are changing current practice guidelines. These recent data, as well as the evidence in favor of using a combination of docetaxel with ADT over ADT alone in the setting of hormone sensitive metastatic PC, are placing ADT alone in a very different therapeutic role as compared to the past. For the purpose of this chapter, therapeutic options in metastatic hormone naïve patients only will be discussed.

1.5.1   Hormonal Therapy Alone

Bilateral orchiectomy, antiandrogens and GnRH agonists have been adopted widely over the last decades, however, the clinical efficacy of the commercially available compounds is hard to compare. Surgical castration in terms of bilateral orchiectomy is considered the gold standard for the ablation of testicular androgens as very low levels of circulating testosterone are reached (≤ 20 ng/dl) [36]. Although medical therapies hardly reach these goals, bilateral orchidectomy has been largely replaced by medical castration. Thus, the present section focuses on historical debates regarding the choice of GnRH alone over combined androgen blockade, as well as the role of bicalutamide monotherapy and GnRH antagonists. Finally, several authors have proposed variation over standard treatment regimens, such as hormonal manipulations and intermittent androgen deprivation, based on the goal to delay castration resistant PC or to reduce side effects related to chronic androgen withdrawal. In most cases these treatments have been adopted outside of practice guidelines.

GnRH alone versus combined androgen blockade. Combined androgen blockade (CAB), consisting of the combination of an AR antagonist plus either a GnRH agonist or bilateral orchiectomy was first introduced in the early 1980s [107]. Since then, a large number of randomized controlled trials have been conducted to evaluate the efficacy of CAB as compared with castration alone. The trials, as well as several meta-analyses, produced contradictory results and failed to provide convincing evidence for CAB [108]. Low statistical power, study immaturity, compliance to treatment, and imbalances in prognostic indicators between study arms of individual trials were implicated as potential sources of discrepancy. Further, cost effectiveness was another point against the combination strategy. More recently, better results in terms of the efficacy and safety of CAB were found when the antiandrogen bicalutamide was used [109]. Finally, choosing between CAB or castration monotherapy remains discretional in this clinical setting, however, bicalutamide 50 mg per day should be considered the antiandrogen of choice in case a combination strategy is preferred.

Bicalutamide monotherapy. Three prospective randomised trials have compared Bicalutamide 150 to ADT in locally advanced and metastatic patients [110, 111]. Bicalutamide monotherapy was as effective as castration in nonmetastatic patients, but there was a small survival advantage for castration in the M1 subgroup [112]. This difference was perhaps partially outweighed by a better tolerability profile and a higher quality of life in patients treated with bicalutamide monotherapy. Higher dose bicalutamide monotherapy (300, 450, and 600 mg) was also tested with regards to tolerability, pharmacokinetics, and clinical efficacy in comparison to castration in M0 and M1 patients [113]. Survival in bicalutamide treated M1 patients was similar to castration-treated M1 patients (differently from previous trials with bicalutamide 150 mg); however, the median PSA level at baseline in patients with M1 disease was <400 ng/ml. Therefore, bicalutamide monotherapy is an option for younger and sexually active patients with locally advanced disease while, for men with a high disease burden (PSA values >400 ng/ml), castration should be considered a better option [114].

Degarelix. As a result of early dose finding phase II and III trials, an initial dose of 240 mg followed by a monthly maintenance dose of 80 mg, were shown to be the best to achieve a rapid, profound, sustained suppression of testosterone and are now used in the clinical setting [43]. In 2008, Klotz reported on CS21, the first phase III randomized controlled trial designed to demonstrate the non-inferiority of degarelix versus leuprolide [9]. The study randomized 610 patients with histologically confirmed PC (all stages), however, the primary end point was testosterone monitoring for 1 year. The survival analysis performed on trial CS21 tended to support an advantage of degarelix on GnRH agonist either in terms of PSA progression free survival or overall survival. However, there were major limitations, including a short follow-up period (1 year) with a limited number of events, mixed inclusion criteria (non-metastatic and metastatic cases) aspect is the follow-up of the trial (only 365 days). In fact, as designed, CS21 could not give answers regarding clinical endpoints, thus, to date we don't have survival data. A meta-analysis was performed focusing on the biochemical and safety profile among the available five trials comparing degarelix to GnRH agonists [115]. According to this work, degarelix produced castration levels in a higher percentage of cases during the first 28 days; however, both treatments were able to maintain testosterone to castration levels to day 364. No significant differences were found regarding PSA level variation. In both groups, adverse events were mild or moderate and dropout rate was comparable and low. The main side effect related to degarelix was site injection reactions. Finally, a subgroup analysis of CS21 generated the hypothesis that degarelix could provide better serum alkaline phosphatase control, compared to leuprolide, mostly in patients with metastatic disease and/or PSA >50 ng/ml [116, 117]; these hypotheses need clinical confirmation.

Intermittent Androgen Deprivation. The concept of intermittent androgen deprivation (IAD) consists of interrupting castration in patients responding to therapy and to restart it later on according to specific PSA driven stop and start rules. This concept was supported by preclinical models showing longer time to hormone resistance with IAD [118]. Further support came from the hypothesis that IAD could reduce treatment costs and improve quality of life (QoL) by reducing side effects seen during continuous ADT, such as compromised sexual functioning, increased risk from cardiovascular diseases and diabetes, osteoporosis, loss of muscle mass, hot flushes, etc. Finally, IAD has found its way to clinics despite weak evidence of its clear superiority or non-inferiority over continuous ADT. Several reviews and meta-analyses have included eight RCTs to look at the clinical efficacy of IAD [119–122]. Among these trials, only three were conducted in patients with exclusively M1 disease, while the remaining included different patient groups, with mixed populations rather than pure cohorts (non-metastatic, metastatic, and locally advanced), and with variable study designs. According to the available data, there is no difference in OS or CSS between IAD and continuous androgen deprivation, however, these trials had limitations, and among those addressing M1 patients only, a trend favoring continuous treatment for OS and PFS was found. In terms of QoL, certain treatment related side effects, such as hot flushes and impairment of sexual activity, were less pronounced with IAD. While the protective effect of IAD over other side effects seen during continuous ADT has been questioned. Finally, patients with large tumors, multiple metastases, and prostate-specific antigen (PSA) levels >100 ng/ml do not have a good prognosis with IAD, mainly due to a shorter life expectancy and a shorter off-treatment period [123].

Hormonal manipulations. Kelly and Scher described for the first time the possibility of secondary hormonal manipulations as a short-term but safe and effective option in patients with failure after primary hormonal therapy. They observed that when an antiandrogen is part of a treatment regimen, discontinuation during biochemical progression resulted in a PSA response in 15–20%, and lasted for 5 months on average [124]. This phenonmenon is called "antiandrogen withdrawal syndrome" (AAWS) and it is likely generated by mutations in the AR gene that enable the antiandrogens to gain function and act as receptor agonists [125]. As a clinical consequence, patients who develop these mutations can benefit from a suspension of antiandrogen treatment. The AAWS has been associated with a longer progression-free survival and greater improvement in quality of life [126]. Further manipulations have been explored, such as switching from one hormonal treatment modality to another. After initial antiandrogen monotherapy, orchiectomy at the time of failure of initial treatment may lead to a PSA response and to symptomatic improvement, as well as second-line treatment with bicalutamide has been shown to improve symptoms and decrease pain in patients without prior antiandrogen therapy. A 50% PSA decrease has been described after second line treatment with non-steroidal antiandrogens in 14–50% of cases [124]. Responders to second-line hormonal treatment may be expected to survive significantly longer than non-responders. However, current guidelines do not mention hormonal manipulations as a potential treatment options. Although hormonal manipulations could be used to postpone the beginning of chemotherapy in metastatic PC patients under CAB, more effective treatment options are now available in this setting.

1.5.2 Hormonal Therapy Combination Regimens

Docetaxel chemotherapy has historically been utilized in the CRPC setting following two randomized clinical trials (RCTs) demonstrating improved overall survival (OS) when compared with mitoxantrone plus prednisone [127, 128].

Similarly, abiraterone acetate demonstrated improved OS in the CRPC setting, both predocetaxel [106, 129] and postdocetaxel treatment [105, 130]. Given the previous positive results, a number of RCTs have more recently demonstrated that the addition of either docetaxel [131–133] or abiraterone [134, 135] to ADT improves OS in men with hormone-naïve metastatic PCa, compared with ADT alone. Similarly, Enzalutamide has improved survival in CRPC patients and is now being tested in hormone sensitive subjects; however, results have not been published up to the present time.

ADT in combination to docetaxel. There has been a long debate regarding the hypothesis of administering early chemotherapy in hormone naive patients, with arguments for and against this approach. In favor is the idea that attacking de novo testosterone independent clones early should allow ADT to keep PC in remission longer. Alternatively, ADT may take cells out of cycle and make them less responsive to cytotoxic treatments. The fact that some patients respond for long periods to ADT and never need chemotherapy is another argument against early chemotherapy. In the past, there have been several clinical reports against early chemotherapy, however it has been noted that none of these trials included cytotoxic therapy shown to prolong overall survival in the setting of metastatic CRPC. The availability of active chemotherapy for CRPC has led to renewed interest and investigation in hormone sensitive disease. In 2013, the results from a French trial, the GETUG-15 trial, were reported in which 385 men with mHSPC were randomized to receive ADT plus docetaxel (75 mg/m^2 every 3 weeks, up to nine cycles) or ADT alone [131]. Although the addition of docetaxel was associated with an improvement in biochemical PFS (p = 0.0021), there was no benefit in OS with the addition of docetaxel, even with long-term follow up (62.1 vs 48.6 months; HR 0.88, 95% CI 0.68–1.14, p = 0.3) [131]. In contrast, the CHAARTED trial randomized 790 men with mHSPC either to ADT or ADT plus docetaxel (75 mg/m^2 every 3 weeks, six cycles) and showed dramatic improvement of OS in the ADT plus docetaxel group than the ADT alone (57.6 vs 44.0 months; HR 0.61, 95% CI 0.47–0.80, p < 0.001) [133]. In parallel, results from the STAMPEDE trial were published, which assigned almost 3000 men with either high-risk localized (24%), node-positive (15%) or mHSPC (61%) to multiple treatment arms. Among patients randomized to receive either ADT alone or ADT plus docetaxel, the addition of docetaxel elicited significant improvement in OS (81.0 vs 71.0 months in ADT plus docetaxel versus ADT alone, respectively; HR 0.78, 95% CI 0.66–0.93, p = 0.006) [132]. The composition of enrolled patients largely differed in the GETUG-15 trial compared with the other two larger studies, which may justify the difference in the clinical results. Although CHAARTED was initially designed as a trial for high-volume metastatic disease, defined as the presence of visceral metastasis and/or four or more osseous metastases, 34.2% of patients with low volume disease were also enrolled. In comparison, the GETUG-15 trial was not initially powered to evaluate this effect of volume of disease. Importantly, among patients with high volume mHSPC, the CHARTEED trial documented an unprecedented 17-month OS improvement with the addition of docetaxel (49.2 vs 32.2 months; HR 0.60, 95% CI 0.45–0.81, p < 0.001), whereas docetaxel originally conferred a relatively modest 2.9-month OS benefit in mCRPC. Further, subgroup analysis of metastatic patients (n = 1087) in STAMPEDE trials also showed substantial improvement of OS for 22 months (65 vs 43 months; HR 0.73, 95% CI 0.59–0.89, p = 0.002). These results have determined a paradigm shift in the first line treatment of patients with metastatic PC, so that most scientific practice guidelines consider ADT plus docetaxel the new standard of care in this setting. Nonetheless, there is no unique definition for the volume of metastatic disease and concerns remains whether or not to consider ADT plus docetaxel the best option for patients with low volume disease. On this issue, a long-term efficacy in the CHAARTED trial was later reported showing that this benefit on OS was not seen in patients with low volume disease [136].

ADT in combination to abiraterone. Two trials have investigated the role of abiraterone compared to ADT alone in the setting of mHSPC (LATITUDE and STAMPEDE arm G) [134, 135]. Another trial, PEACE-1, is still actively recruiting. The data from LATITUDE and STAMPEDE reported that combination use of abiraterone with ADT also offers improvement of OS compared with ADT alone in mHSPC. These data showed the robust effect of abiraterone, but also raise the controversial discussion for the optimal sequence or combination use with regards to docetaxel and to other effective compounds. The STAMPEDE arm G cohort involved patients with non-metastatic high-risk disease (characterized by T3 or T4, tumor-positive lymph nodes; Gleason's score of 8–10 and/or serum PSA level of ≥40 ng/ml): 915 and 1002 patients in non-metastatic and metastatic groups, respectively. The advantage on OS of combination use of abiraterone was only observed in M1 patients, not in M0 patients. The LATITUDE trial enrolled only M1 patients, and showed a significant improvement on OS by combination use of abiraterone with ADT. Interestingly, daily abiraterone 1000 mg was associated to a lower dose of prednisone compared to the CRPC regimen (5 mg by mouth daily versus 5 mg twice daily). A limitation of this trial is that only 27% of men in the ADT only arm received abiraterone or enzalutamide at progression to mCRPC, and only 52% of these men received any life-prolonging therapy. A systematic review of the M1 cohort in STAMPEDE arm G (1002 patients) and LATITUDE (1199 patients) reported that survival benefit might be greater in younger patients (aged <75 years) despite the small sample size of older patients (137/1002 and 202/1199 in STAMPEDE arm G and LATITUDE, respectively) [137]. Unfortunately, these two studies did not define low and high tumor volume in their analyses, therefore, whether this survival benefit with the combination of abiraterone is delivered in lower tumor burden patients remains unknown. Nevertheless, abiraterone should be considered an option in addition to ADT for men with mHSPC.

Conclusion. Based on the available data, docetaxel plus ADT, and abiraterone plus prednisone plus ADT, are both standard of care treatment options for the management of high volume mHSPC, however abiraterone has lower toxicity. Among patients with lower treatment burden, abiraterone plus prednisone plus ADT could be considered a preferred standard of care due to a lack of a clear benefit with docetaxel in this subgroup.

1.6 Side Effects of Hormonal Therapy

In many respects, the role of hormonal therapy in PC has overshadowed its impact on toxicity and QoL. When ADT was introduced as a treatment for metastatic PC, bothersome side effects were an acceptable trade-off for managing symptoms and improving life expectancy. However, ever since indications for ADT have expanded improvements in life expectancy have been uncertain, particularly for lower-grade cancer [138]. Recently, the body of literature highlighting toxicities from hormonal therapy has mounted, and includes increased risk of cardiovascular disease (CVD), harmful metabolic changes such as obesity, insulin resistance and diabetes, dyslipidemia, and changes in bone health leading to a higher risk of bone fracture.

1.6.1 Cardiovascular Disease

Several observational/retrospective studies and secondary analyses from randomized trials have found an association between long term ADT and the risk of developing either nonfatal or fatal CVD [139–142]. Short term ADT (4 months) may be fatal only in patients with a history of congestive heart failure or myocardial infarction [143]. In these patients, revascularization prior to the start of shortterm ADT may reduce, but not eliminate, the 5-year overall mortality risk [144]. Among all endocrine treatment modalities analyzed, surgical castration and the use of antiandrogen monotherapy seemed to have a lower impact on CVD [139, 140, 142]. Although several analyses in the literature suggest a better cardiovascular safety profile for degarelix [145, 146], a meta-analysis found that severe cardiovascular side effects were lower in the degarelix group, but not significantly [115]. Conversely, a large study from a Canadian database found that neither the use nor the duration of ADT was associated with an increased risk of acute myocardial infarction or sudden cardiac death [147].

Although data on ADT related risk for cardiovascular events and mortality are inconsistent, the use of the American Heart Association guidelines could be applied. Primary prevention should feature tobacco cessation and management of hypertension. Low-dose aspirin is recommended for men with a 10% or greater 10 year risk for coronary heart disease. Lifestyle modification should feature weight control and low intake of saturated fat and cholesterol. If such modifications fail to achieve target LDL, statins should be used as first-line drug treatment of hyperlipidemia [148].

1.6.2 Metabolic Changes

The term sarcopenic obesity describes the increase of abdominal obesity accompanied by reduced muscle mass in men undergoing ADT [149], and is found to be associated with an increase in all-cause mortality [150]. Several studies have shown that in men on ADT, body weight and body composition changes occur mainly during the first year of treatment and subsequently continue for 1–2 years [151]. In response to ADT, patients with PC may experience an increase in fasting insulin and a decrease in insulin receptor sensitivity [152, 153]. Although randomized trials are lacking, observational studies have also documented an increased risk of incident diabetes secondary to ADT, possibly related to time of exposure [139–141, 147]. As the risk for diabetes development during ADT is high, the need for diabetes screening among men with PC under long-term treatment is mandatory.

1.6.3 Changes in Bone Health

Androgen deprivation may progressively cause a decrease in bone mineral density and an increase in the incidence of osteoporosis or fracture. This prevalence of osteoporosis is variable, 9–53%, and largely depending on treatment duration, disease stage, ethnicity and site of osteoporosis measurement [154]. Fracture risk was shown to be highest in long-term users of GnRH agonist and in men undergoing bilateral orchiectomy [155], while bicalutamide, when compared with leuprolide, had a lower impact on bone metabolism disorders [156]. Fractures associated with ADT may result in hospitalization and finally cause an increase in mortality [157]. In patients at higher risk, preventive measures should be taken into account. Denosumab has shown to increase the lumbar bone mineral density by 5.6% and reduce vertebral fracture using a 60-mg regimen every 6 months [158]. Zoledronic acid has also shown to increase bone mineral density, although an optimal regimen for the prevention of skeletal-related events in a hormone-sensitive PC has not been determined [159].

References

1. Huggins C, Hodges CV (1941) Studies on prostatic cancer II: the effects of castration on advanced carcinoma of the prostate gland. Arch Surg 43:209–223
2. Trachtenberg J (1987) Hormonal management of stage D carcinoma of the prostate. Urol Clin North Am 14(4):685–694
3. The Veterans Administration Co-Operative Urological Research Group (1967) Treatment and survival of patients with cancer of the prostate. Surg Gynecol Obstet 124:1011–1017
4. Byar DP (1973) The Veterans Administration Cooperative Research Group's studies of cancer of the prostate. Cancer 32:1126–1130
5. Byar DP, Corle DK (1988) Hormone therapy for prostate cancer: results of the Veterans Administration Cooperative Urological Research Group studies. NCI Monogr 43:209–223
6. Schally AV, Arimura A, Baba Y et al (1971) Isolation and properties of the FSH and LH-releasing hormone. Biochem Biophys Res Commun 43:393–399
7. The Leuprolide Study Group (1984) Leuprolide versus diethylstilbestrol for metastatic prostate cancer. N Engl J Med 311:1281–1286
8. Ferring Pharmaceuticals (2008) FDA approves Ferring Pharmaceuticals' degarelix (generic name) for treatment of advanced prostate cancer [press release], 4 Dec 2008

9. Klotz L, Boccon-Gibod L, Shore ND et al (2008) The efficacy and safety of degarelix: a 12-month, comparative, randomized, open-label, parallel-group phase III study in patients with prostate cancer. BJU Int 102(11):1531–1538

10. Labrie F, Luu-The V, Bélanger A et al (2005) Is dehydroepiandrosterone a hormone? J Endocrinol 187(2):169–196

11. Labrie F, Dupont A, and Belanger A. (1985) Complete androgen blockade for the treatment of prostate cancer. In: de Vita VT, Hellman S, Rosenberg SA (eds) Important Advances in Oncology, Philadelphia: J.B. Lippincott, pp 193–217

12. Belanger B, Belanger A, Labrie F et al (1989) Comparison of residual C-19 steroids in plasma and prostatic tissue of human, rat and guinea pig after castration: unique importance of extratesticular androgens in men. J Steroid Biochem 32:695–698

13. Miyamoto H, Yeh S, Lardy H, Messing E, Chang C (1998) Delta 5-androstenediol is a natural hormone with androgenic activity in human prostate cancer cells. Proc Natl Acad Sci U S A 95:11083–11088

14. Mostaghel EA, Nelson PS, Lange P et al (2014) Targeted androgen pathway suppression in localized prostate cancer: a pilot study. J Clin Oncol 32:229–237

15. Heemers HV, Tindall DJ (2007) Androgen receptor (AR) coregulators: a diversity of functions converging on and regulating the AR transcriptional complex. Endocr Rev 28:778–808

16. Robinson-Rechavi M, Escriva Garcia H, Laudet V (2003) The nuclear receptor superfamily. J Cell Sci 116:585–586

17. Ruizeveld De Winter JA, Trapman J et al (1991) Androgen receptor expression in human tissues: an immunohistochemical study. J Histochem Cytochem 39(7):927–936

18. Evangelou AI, Winter SF, Huss WJ, Bok RA, Greenberg NM (2004) Steroid hormones, polypeptide growth factors, hormone refractory prostate cancer, and the neuroendocrine phenotype. J Cell Biochem 91(4):671–683

19. Hayward S, Rosen M, Cunha G (1997) Stromal-epithelial interactions in the normal and neoplastic prostate. Br J Urol 79(S2):18–26

20. Wang Y, Hayward SW, Cao M, Thayer KA, Cunha GR (2001) Cell differentiation lineage in the prostate. Differentiation 68(4–5):270–279

21. Kwabi-Addo B, Ozen M, Ittmann M (2004) The role of fibroblast growth factors and their receptors in prostate cancer. Endocr Relat Cancer 11(4):709–724

22. Levine AC, Liu XH, Greenberg PD et al (1988) Androgens induce the expression of vascular endothelial growth factor in human fetal prostatic fibroblasts. Endocrinology 139(11):4672–4678

23. Yu S, Yeh CR, Niu Y et al (2012) Altered prostate epithelial development in mice lacking the androgen receptor in stromal fibroblasts. Prostate 72(4):437–449

24. Wu CT, Altuwaijri S, Ricke WA et al (2007) Increased prostate cell proliferation and loss of cell differentiation in mice lacking prostate epithelial androgen receptor. PNAS 104:12679–12684

25. Zhou Y, Bolton EC, Jones JO (2015) Androgens and androgen receptor signaling in prostate tumorigenesis. J Mol Endocrinol 54(1):R15–R29

26. Imamoto T, Suzuki H, Fukasawa S et al (2005) Pretreatment serum testosterone level as a predictive factor of pathological stage in localized prostate cancer patients treated with radical prostatectomy. Eur Urol 47:308–312

27. Stattin P, Lumme S, Tenkanen L et al (2004) High levels of circulating testosterone are not associated with increased prostate cancer risk: a pooled prospective study. Int J Cancer 108:418–424

28. Andriole G, Bostwick D, Brawley O et al (2004) Chemoprevention of prostate cancer in men at high risk: rationale and design of the reduction by dutasteride of prostate cancer events (REDUCE) trial. J Urol 172:1314–1317

29. Goodman PJ, Tangen CM, Crowley JJ et al (2004) Implementation of the Prostate Cancer Prevention Trial (PCPT). Control Clin Trials 25:203–222

30. Kyprianou N, Isaacs JT (1988) Activation of programmed cell death in the rat ventral prostate after castration. Endocrinology 122:552–562

31. Reuter VE (1997) Pathological changes in benign and malignant prostatic tissue following androgen deprivation therapy. Urology 49:16–22
32. Westin P, Stattin P, Damber JE, Bergh A (1995) Castration therapy rapidly induces apoptosis in a minority and decreases cell proliferation in a majority of human prostatic tumors. Am J Pathol 146:1368–1375
33. Rosner W, Auchus RJ, Azziz R, Sluss PM, Raff H (2007) Position statement: utility, limitations, and pitfalls in measuring testosterone: an Endocrine Society position statement. J Clin Endocrinol Metab 92(2):405–413
34. Bhasin S, Cunningham GR, Hayes FJ et al (2010) Testosterone therapy in men with androgen deficiency syndromes: an Endocrine Society clinical practice guideline. J Clin Endocrinol Metab 95(6):2536–2559
35. Bubley GJ, Carducci M, Dahut W et al (1999) Eligibility and response guidelines for phase II clinical trials in androgen-independent prostate cancer: recommendations from the Prostate-Specific Antigen Working Group. J Clin Oncol 17(11):3461–3467
36. Oefelein MG, Feng A, Scolieri MJ, Ricchiutti D, Resnick MI (2000) Reassessment of the definition of castrate levels of testosterone: implications for clinical decision making. Urology 56:1021–1024
37. Potosky AL, Knopf K, Clegg LX et al (2001) Quality-of-life outcomes after primary androgen deprivation therapy: results from the Prostate Cancer Outcomes Study. J Clin Oncol 19(17):3750–3757
38. Seidenfeld J, Samson DJ, Hasselblad V et al (2000) Single-therapy androgen suppression in men with advanced prostate cancer: a systematic review and meta-analysis. Ann Intern Med 132(7):566–577
39. Bubley GJ (2001) Is the flare phenomenon clinically significant? Urology 58(2 Suppl 1):5–9
40. Labrie F, Dupont A, Belanger A et al (1987) Flutamide eliminates the risk of disease flare in prostatic cancer patients treated with a luteinizing hormone-releasing hormone agonist. J Urol 138:804–806
41. Nishiyama T (2014) Serum testosterone levels after medical or surgical androgen deprivation: a comprehensive review of the literature. Urol Oncol 32(1):38.e17–38.e28
42. Morote J, Planas J, Salvador C et al (2009) Individual variations of serum testosterone in patients with prostate cancer receiving androgen deprivation therapy. BJU Int 103(3):332–335
43. Van Poppel H, Tombal B, de la Rosette JJ et al (2008) Degarelix: a novel gonadotropin-releasing hormone (GnRH) receptor blocker—results from a 1-yr, multicentre, randomised, phase 2 dosage-finding study in the treatment of prostate cancer. Eur Urol 54:805–813
44. Mahler C, Verhelst J, Denis L (1988) Clinical pharmacokinetics of the antiandrogens and their efficacy in prostate cancer. Clin Pharmacokinet 34(5):405–417
45. Barrie SE, Haynes BP, Potter GA et al (1997) Biochemistry and pharmacokinetics of potent non-steroidal cytochrome P450 (17alpha) inhibitors. J Steroid Biochem Mol Biol 60:347–351
46. Attard G, Reid AH, Auchus RJ et al (2012) Clinical and biochemical consequences of CYP17A1 inhibition with abiraterone given with and without exogenous glucocorticoids in castrate men with advanced prostate cancer. J Clin Endocrinol Metab 97:507–516
47. Tran C, Ouk S, Clegg NJ et al (2009) Development of a second-generation antiandrogen for treatment of advanced prostate cancer. Science 324(5928):787–790
48. Guerrero J, Alfaro IE, Gomez F, Protter AA, Bernales S (2013) Enzalutamide, an androgen receptor signaling inhibitor, induces tumor regression in a mouse model of castration-resistant prostate cancer. Prostate 73:1291–1305
49. Scher HI, Fizazi K, Saad F et al (2012) Increased survival with enzalutamide in prostate cancer after chemotherapy. N Engl J Med 367(13):1187–1197
50. Mozayani A, Raymon L (2011) Handbook of drug interactions: a clinical and forensic guide. Springer, New York, NY, p 656
51. Grino PB, Griffin JE, Wilson JD (1990) Testosterone at high concentrations interacts with the human androgen receptor similarly to dihydrotestosterone. Endocrinology 126(2):1165–1172
52. Iczkowski KA, Qiu J, Qian J et al (2005) The dual 5-alpha-reductase inhibitor dutasteride induces atrophic changes and decreases relative cancer volume in human prostate. Urology 65:76–82

53. Mariani S, Salvatori L, Basciani S et al (2006) Expression and cellular localization of follicle-stimulating hormone receptor in normal human prostate, benign prostatic hyperplasia and prostate cancer. J Urol 175:2072–2077
54. Radu A, Pichon C, Camparo P et al (2010) Expression of follicle-stimulating hormone receptor in tumor blood vessels. N Engl J Med 363:1621–1630
55. Kaku H, Saika T, Tsushima T et al (2006) Time course of serum testosterone and luteinizing hormone levels after cessation of long-term luteinizing hormone-releasing hormone agonist treatment in patients with prostate cancer. Prostate 66:439–444
56. Wilke DR, Parker C, Andonowski A et al (2006) Testosterone and erectile function recovery after radiotherapy and long-term androgen deprivation with luteinizing hormone-releasing hormone agonists. BJU Int 97:963–968
57. Morley JE, Kaiser FE, Perry HM 3rd et al (1997) Longitudinal changes in testosterone, luteinizing hormone, and follicle-stimulating hormone in healthy older men. Metabolism 46:410–413
58. Nejat RJ, Rashid HH, Bagiella E, Katz AE, Benson MC (2000) A prospective analysis of time to normalization of serum testosterone after withdrawal of androgen deprivation therapy. J Urol 164:1891–1894
59. Yoon FH, Gardner SL, Danjoux C et al (2008) Testosterone recovery after prolonged androgen suppression in patients with prostate cancer. J Urol 180:1438–1444
60. Gulley JL, Figg WD, Steinberg SM et al (2005) A prospective analysis of the time to normalization of serum androgens following 6 months of androgen deprivation therapy in patients on a randomized phase III clinical trial using limited hormonal therapy. J Urol 173(5):1567–1571
61. Pettersson B, Varenhorst E, Petas A, Sandow J (2006) Duration of testosterone suppression after a 9.45 mg implant of the GnRH-analogue buserelin in patients with localized carcinoma of the prostate a 12-month follow-up study. Eur Urol 50:483–489
62. Pickles T, Agranovich A, Berthelet E et al (2002) Testosterone recovery following prolonged adjuvant androgen ablation for prostate carcinoma. Cancer 94:362–367
63. Cooperberg MR, Grossfeld GD, Lubeck DP et al (2003) National practice patterns and time trends in androgen ablation for localized prostate cancer. J Natl Cancer Inst 95:981–989
64. Kawakami J, Cowan JE, Elkin EP et al (2006) Androgen-deprivation therapy as primary treatment for localized prostate cancer: data from Cancer of the Prostate Strategic Urologic Research Endeavor (CaPSURE). Cancer 106:1708–1714
65. Lu-Yao GL, Albertsen PC, Moore DF et al (2008) Survival following primary androgen deprivation therapy among men with localized prostate cancer. JAMA 300:173–181
66. Wong YN, Freedland SJ, Egleston B et al (2009a) Role of androgen deprivation therapy for node-positive prostate cancer. J Clin Oncol 27:100–105
67. Wong YN, Freedland SJ, Egleston B et al (2009b) The role of primary androgen deprivation therapy in localized prostate cancer. Eur Urol 56:609–616
68. Studer UE, Whelan P, Albrecht W et al (2006) Immediate or deferred androgen deprivation for patients with prostate cancer not suitable for local treatment with curative intent: European Organisation for Research and Treatment of Cancer (EORTC) Trial 30891. J Clin Oncol 24:1868–1876
69. McLeod DG, Iversen P, See WA et al (2006) Bicalutamide 150 mg plus standard care vs standard care alone for early prostate cancer. BJU Int 97:247–254
70. Mason MD, Parulekar WR, Sydes MR et al (2015) Final report of the intergroup randomized study of combined androgen-deprivation therapy plus radiotherapy versus androgen-deprivation therapy alone in locally advanced prostate cancer. J Clin Oncol 33:2143–2150
71. Mottet N, Peneau M, Mazeron JJ, Molinie V, Richaud P (2012) Addition of radiotherapy to long-term androgen deprivation in locally advanced prostate cancer: an open randomised phase 3 trial. Eur Urol 62(2):213–219
72. Widmark A, Klepp O, Solberg A et al (2009) Endocrine treatment, with or without radiotherapy, in locally advanced prostate cancer (SPCG-7/SFUO-3): an open randomized phase III trial. Lancet 373:301–308

73. Engel J, Bastian PJ, Baur H et al (2010) Survival benefit of radical prostatectomy in lymph node–positive patients with prostate cancer. Eur Urol 57:754–761
74. Ghavamian R, Bergstralh EJ, Blute ML, Slezak J, Zincke H (1999) Radical retropubic prostatectomy plus orchiectomy versus orchiectomy alone for pTxN+ prostate cancer: a matched comparison. J Urol 161:1223–1227
75. Grimm M-O, Kamphausen S, Hugenschmidt H et al (2002) Clinical outcome of patients with lymph node positive prostate cancer after radical prostatectomy versus androgen deprivation. Eur Urol 41:628–634
76. Kaminski JM, Hanlon AL, Joon DL et al (2003) Effect of sequencing of androgen deprivation and radiotherapy on prostate cancer growth. Int J Radiat Oncol Biol Phys 57:24–28
77. Milosevic M, Chung P, Parker C et al (2007) Androgen withdrawal in patients reduces prostate cancer hypoxia: implications for disease progression and radiation response. Cancer Res 67(13):6022–6025
78. Lu JP, Monardo L, Bryskin I et al (2010) Androgens induce oxidative stress and radiation resistance in prostate cancer cells though NADPH oxidase. Prostate Cancer Prostatic Dis 13(1):39–46
79. D'Amico AV, Chen MH, Renshaw AA, Loffredo M, Kantoff PW (2008) Androgen suppression and radiation vs radiation alone for prostate cancer, a randomized trial. JAMA 299:289–295
80. Denham JW, Steigler A, Lamb DS et al (2005) Short-term androgen deprivation and radiotherapy for locally advanced prostate cancer: results from the Trans-Tasman Radiation Oncology Group 96.01 randomized controlled trial. Lancet Oncol 6:841–850
81. Horwitz EM, Bae K, Hanks GE et al (2008) Ten-year follow-up of Radiation Therapy Oncology Group Protocol 92-02: a phase III trial of the duration of elective androgen deprivation in locally advanced prostate cancer. J Clin Oncol 26:2497–2504
82. Roach M, Bae K, Speight J et al (2008) Short-term neoadjuvant androgen deprivation therapy and external-beam radiotherapy for locally advanced prostate cancer: long-term results of RTOG 8610. J Clin Oncol 26:585–591
83. Zeliadt SB, Potosky AL, Penson DF, Etzioni R (2006) Survival benefit associated with adjuvant androgen deprivation therapy combined with radiotherapy for high- and low-risk patients with nonmetastatic prostate cancer. Int J Radiat Oncol Biol Phys 66(2):395–402
84. Ciezki JP, Klein EA, Angermeier K et al (2004) A retrospective comparison of androgen deprivation (AD) vs. no AD among low-risk and intermediate-risk prostate cancer patients treated with brachytherapy, external beam radiotherapy, or radical prostatectomy. Int J Radiat Oncol Biol Phys 60:1347–1350
85. D'Amico AV, Loffredo M, Renshaw AA, Loffredo B, Chen MH (2006) Six-month androgen suppression plus radiation therapy compared with radiation therapy alone for men with prostate cancer and a rapidly increasing pretreatment prostate-specific antigen level. J Clin Oncol 24(25):4190–4195
86. Jones CU, Hunt D, McGowan DG et al (2011) Radiotherapy and shortterm androgen deprivation for localized prostate cancer. N Engl J Med 365:107–118
87. Pisansky TM, Hunt D, Gomella LG et al (2015) Duration of androgen suppression before radiotherapy for localized prostate cancer: radiation therapy oncology group randomized clinical trial 9910. J Clin Oncol 33:332–339
88. Bolla M, Collette L, Blank L et al (2002) Long-term results with immediate androgen suppression and external irradiation in patients with locally advanced prostate cancer (an EORTC study): a phase III randomised trial. Lancet 360:103–106
89. Bolla M, de Reijke TM, Van Tienhoven G et al (2009) Duration of androgen suppression in the treatment of prostate cancer. N Engl J Med 360:2516–2527
90. Freedland SJ, Humphreys EB, Mangold LA et al (2005) Risk of prostate cancer-specific mortality following biochemical recurrence after radical prostatectomy. JAMA 294:433–439
91. Makarov DV, Humphreys EB, Mangold LA et al (2008) The natural history of men treated with deferred androgen deprivation therapy in whom metastatic prostate cancer developed following radical prostatectomy. J Urol 179:156–161

92. Messing EM, Manola J, Sarosdy M et al (1999) Immediate hormonal therapy compared with observation after radical prostatectomy and pelvic lymphadenectomy in men with node-positive prostate cancer. N Engl J Med 341:1781–1788

93. Eisenberger MA, Walsh PC (1999) Early androgen deprivation for prostate cancer? N Engl J Med 341:1837

94. Moul JW, Wu H, Sun L et al (2004) Early versus delayed hormonal therapy for prostate specific antigen only recurrence of prostate cancer after radical prostatectomy. J Urol 171:1141–1147

95. Siddiqui SA, Boorjian SA, Inman B et al (2008) Timing of androgen deprivation therapy and its impact on survival after radical prostatectomy: a matched cohort study. J Urol 179:1830–1837

96. Andriole G, Lieber M, Smith J et al (1995) Treatment with finasteride following radical prostatectomy for prostate cancer. Urology 45:491–497

97. Schröder F, Bangma C, Angulo JC et al (2013) Dutasteride treatment over 2 years delays prostate-specific antigen progression in patients with biochemical failure after radical therapy for prostate cancer: results from the randomised, placebo-controlled Avodart After Radical Therapy for Prostate Cancer Study (ARTS). Eur Urol 63:779–787

98. Shin YS, Lee JW, Kim MK, Jeong YB, Park SC (2017) Early dutasteride monotherapy in men with detectable serum prostate-specific antigen levels following radical prostatectomy: a prospective trial. Investig Clin Urol 58(2):98–102

99. Weiner AB, Matulewicz RS, Eggener SE et al (2016) Increasing incidence of metastatic prostate cancer in the United States (2004 2013). Prostate Cancer Prostatic Dis 19:395–397

100. Harris WP, Mostaghel EA, Nelson PS, Montgomery B (2009) Androgen deprivation therapy: progress in understanding mechanisms of resistance and optimizing androgen depletion. Nat Clin Pract Urol 6:76–85

101. West TA, Kiely BE, Stockler MR (2014) Estimating scenarios for survival time in men starting systemic therapies for castration-resistant prostate cancer: a systematic review of randomised trials. Eur J Cancer 50:1916–1924

102. Mostaghel EA, Nelson PS (2008) Intracrine androgen metabolism in prostate cancer progression: mechanisms of castration resistance and therapeutic implications. Best Pract Res Clin Endocrinol Metab 22:243–258

103. Mostaghel EA, Page ST, Lin DW et al (2007) Intraprostatic androgens and androgen-regulated gene expression persist after testosterone suppression: therapeutic implications for castration-resistant prostate cancer. Cancer Res 67:5033–5041

104. Beer TM, Armstrong AJ, Rathkopf DE et al (2014) Enzalutamide in metastatic prostate cancer before chemotherapy. N Engl J Med 371(5):424–433

105. de Bono JS, Logothetis CJ, Molina A et al (2011) Abiraterone and increased survival in metastatic prostate cancer. N Engl J Med 364:1995–2005

106. Ryan CJ, Smith MR, de Bono JS et al (2013) Abiraterone in metastatic prostate cancer without previous chemotherapy. N Engl J Med 368:138–148

107. Labrie F, Dupont A, Belanger A et al (1982) New hormonal therapy in prostatic carcinoma: combined treatment with an LHRH agonist and an antiandrogen. Clin Invest Med 5:267–275

108. Pagliarulo V, Bracarda S, Eisenberger MA et al (2012) Contemporary role of androgen deprivation therapy for prostate cancer. Eur Urol 61(1):11–25

109. Akaza H, Hinotsu S, Usami M et al (2009) Study group for the combined androgen blockade therapy of prostate cancer combined androgen blockade with bicalutamide for advanced prostate cancer: long-term follow-up of a phase 3, double-blind, randomized study for survival. Cancer 115:3437–3445

110. Boccardo F, Barichello M, Battaglia M et al (2002) Bicalutamide monotherapy versus flutamide plus goserelin in prostate cancer: updated results of a multicentric trial. Eur Urol 42:481–490

111. Iversen P, Tyrrell CJ, Kaisary AV et al (2000) Bicalutamide monotherapy compared with castration in patients with nonmetastatic locally advanced prostate cancer: 6.3 years of follow-up. J Urol 164:1579–1582

112. Tyrrell CJ, Kaisary AV, Iversen P et al (1998) A randomised comparison of 'Casodex'™ (bicalutamide) 150 mg monotherapy versus castration in the treatment of metastatic and locally advanced prostate cancer. Eur Urol 33:447–456

113. Tyrrell CJ, Iversen P, Tammela T et al (2006) Tolerability, efficacy and pharmacokinetics of bicalutamide 300 mg, 450 mg or 600 mg as monotherapy for patients with locally advanced or metastatic prostate cancer, compared with castration. BJU Int 98:563–572

114. Kaisary AV, Iversen P, Tyrrell CJ, Carroll K, Morris T (2011) Is there a role for antiandrogen monotherapy in patients with metastatic prostate cancer? Prostate Cancer Prostatic Dis 4:196–203

115. Sciarra A, Fasulo A, Ciardi A et al (2016) A meta-analysis and systematic review of randomized controlled trials with degarelix versus gonadotropin-releasing hormone agonists for advanced prostate cancer. Medicine (Baltimore) 95(27):e3845

116. Schröder FH, Tombal B, Miller K et al (2010) Changes in alkaline phosphatase levels in patients with prostate cancer receiving degarelix or leuprolide: results from a 12-month, comparative, phase III study. BJU Int 106(2):182–187

117. Tombal B, Miller K, Boccon-Gibod L et al (2010) Additional analysis of the secondary end point of biochemical recurrence rate in a phase 3 trial (CS21) comparing Degarelix 80mg versus leuprolide in prostate cancer patients segmented by baseline characteristics. Eur Urol 57:836–842

118. Sato N, Gleave ME, Bruchovsky N et al (1996) Intermittent androgen suppression delays progression to androgen-independent regulation of prostate-specific antigen gene in the LNCaP prostate tumour model. J Steroid Biochem Mol Biol 58:139–146

119. Brungs D, Chen J, Masson P, Epstein RJ (2014) Intermittent androgen deprivation is a rational standard-of-care treatment for all stages of progressive prostate cancer: results from a systematic review and meta-analysis. Prostate Cancer Prostatic Dis 17:105–111

120. Hussain M, Tangen CM, Berry DL et al (2013) Intermittent versus continuous androgen deprivation in prostate cancer. N Engl J Med 368:1314–1325

121. Magnan S, Zarychanski R, Pilote L et al (2015) Intermittent vs continuous androgen deprivation therapy for prostate cancer: a systematic review and meta-analysis. JAMA Oncol 1:1261–1269

122. Niraula S, Le LW, Tannock IF (2013) Treatment of prostate cancer with intermittent versus continuous androgen deprivation: a systematic review of randomized trials. J Clin Oncol 31(16):2029–2036

123. Hussain M, Tangen C, Higano C, Vogelzang N, Thompson I (2016) Evaluating intermittent androgen deprivation therapy phase III clinical trials: the devil is in the details. J Clin Oncol 34:280–285

124. Kelly WK, Scher HI (1993) Prostate specific antigen decline after antiandrogen withdrawal: the flutamide withdrawal syndrome. J Urol 149:607–609

125. Hara T, Miyazaki J, Araki H (2003) Novel mutations of androgen receptor: a possible mechanism of bicalutamide withdrawal syndrome novel mutations of androgen receptor. Cancer Res 63:149–153

126. Dupont A, Gomez JL, Cusan L et al (1993) Response to flutamide withdrawal in advanced prostate cancer in progression under combination therapy. J Urol 150:908–913

127. Petrylak DP, Tangen CM, Hussain MH et al (2004) Docetaxel and estramustine compared with mitoxantrone and prednisone for advanced refractory prostate cancer. N Engl J Med 351:1513–1520

128. Tannock IF, de Wit R, Berry WR et al (2004) Docetaxel plus prednisone or mitoxantrone plus prednisone for advanced prostate cancer. N Engl J Med 351:1502–1512

129. Rathkopf DE, Smith MR, de Bono JS et al (2014) Updated interim efficacy analysis and long-term safety of abiraterone acetate in metastatic castration-resistant prostate cancer patients without prior chemotherapy (COU-AA-302). Eur Urol 66:815–825

130. Fizazi K, Scher HI, Molina A et al (2012) Abiraterone acetate for treatment of metastatic castration-resistant prostate cancer: final overall survival analysis of the COU-AA-301 randomised, double-blind, placebo-controlled phase 3 study. Lancet Oncol 13:983–992

131. Gravis G, Boher JM, Joly F et al (2016) Androgen deprivation therapy (ADT) plus docetaxel versus ADT alone in metastatic non castrate prostate cancer: impact of metastatic burden and long-term survival analysis of the randomized phase 3 GETUG-AFU15 Trial. Eur Urol 70:256–262
132. James ND, Sydes MR, Clarke NW et al (2016) Addition of docetaxel, zoledronic acid, or both to first-line long-term hormone therapy in prostate cancer (STAMPEDE): survival results from an adaptive, multiarm, multistage, platform randomised controlled trial. Lancet 387:1163–1177
133. Sweeney CJ, Chen YH, Carducci M et al (2015) Chemohormonal therapy in metastatic hormone-sensitive prostate cancer. N Engl J Med 373:737–746
134. Fizazi K, Tran N, Fein L et al (2017) Abiraterone plus prednisone in metastatic, castration-sensitive prostate cancer. N Engl J Med 377:352–360
135. James ND, de Bono JS, Spears MR et al (2017) Abiraterone for prostate cancer not previously treated with hormone therapy. N Engl J Med 377:338–351
136. Sweeney C, Chen YH, Liu G et al (2016) Long term efficacy and QOL data of chemohormonal therapy (C-HT) in low and high volume hormone naïve metastatic prostate cancer (PrCa): E3805 CHAARTED trial. Ann Oncol 27:720PD-PD
137. Rydzewska LHM, Burdett S, Vale CL et al (2017) Adding abiraterone to androgen deprivation therapy in men with metastatic hormone-sensitive prostate cancer: a systematic review and meta-analysis. Eur J Cancer 84:88–101
138. Loblaw DA, Virgo KS, Nam R et al (2007) Initial hormonal management of androgen-sensitive metastatic, recurrent, or progressive prostate cancer: 2006 update of an American Society of Clinical Oncology practice guideline. J Clin Oncol 25:1596–1605
139. Keating NJ, O'Malley AJ, Smith MR (2006) Diabetes and cardiovascular disease during androgen deprivation therapy for prostate cancer. J Clin Oncol 24:4448–4456
140. Keating NL, O'Malley AJ, Freedland SJ, Smith MR (2010) Diabetes and cardiovascular disease during androgen deprivation therapy: observational study of veterans with prostate cancer. J Natl Cancer Inst 102:39 46
141. Tsai HK, D'Amico AV, Sadetsky N, Chen MH, Carroll PR (2007) Androgen deprivation therapy for localized prostate cancer and the risk of cardiovascular mortality. J Natl Cancer Inst 99:1516–1524
142. Van Hemelrijck M, Garmo H, Holmberg L et al (2010) Absolute and relative risk of cardiovascular disease in men with prostate cancer: results from the Population-Based PCBaSe Sweden. J Clin Oncol 28:3448–3456
143. Nanda AN, Chen MH, Braccioforte MH, Moran BJ, D'Amico AV (2009) Hormonal therapy use for prostate cancer and mortality in men with coronary artery disease-induced congestive heart failure or myocardial infarction. JAMA 302:866–873
144. Nguyen PL, Chen MH, Goldhaber SZ et al (2011) Coronary revascularization and mortality in men with congestive heart failure or prior myocardial infarction who receive androgen deprivation. Cancer 117:406–413
145. Albertsen PC, Klotz L, Tombal B et al (2014) Cardiovascular morbidity associated with gonadotropin releasing hormone agonists and an antagonist. Eur Urol 65:565–573
146. Smith MR, Klotz L, van der Meulen E et al (2011) Gonadotropin-releasing hormone blockers and cardiovascular disease risk: analysis of prospective clinical trials of degarelix. J Urol 186:1835–1842
147. Alibhai SM, Duong-Hua M, Sutradhar R et al (2009) Impact of androgen deprivation therapy on cardiovascular disease and diabetes. J Clin Oncol 27:3452–3458
148. Lim TH, Orija IB, Pearlman BL (2014) American College of Cardiology; American College of Cardiology. The new cholesterol treatment guidelines from the American College of Cardiology/American Heart Association, 2013: what clinicians need to know. Postgrad Med 126:35–44
149. Zamboni M, Mazzali G, Fantin F, Rossi A, Di Francesco V (2000) Sarcopenic obesity: a new category of obesity in the elderly. Nutr Metab Cardiovasc Dis 18:388–395

150. Tian S, Xu Y (2016) Association of sarcopenic obesity with the risk of all-cause mortality: a meta-analysis of prospective cohort studies. Geriatr Gerontol Int 16:155–166
151. Smith MR, Lee H, Fallon M, Nathan DM (2008) Adipocytokines, obesity and insulin resistance during combined androgen blockade for prostate cancer. Urology 71:318–322
152. Smith JC, Bennett S, Evans LM et al (2001) The effects of induced hypogonadism on arterial stiffness, body composition, and metabolic parameters in males with prostate cancer. J Clin Endocrinol Metab 86:4261–4267
153. Smith MR, Lee H, Nathan DM (2006) Insulin sensitivity during combined androgen blockade for prostate cancer. J Clin Endocrinol Metab 91:1305–1308
154. Lassemillante AC, Doi SA, Hooper JD, Prins JB, Wright OR (2014) Prevalence of osteoporosis in prostate cancer survivors: a meta-analysis. Endocrine 45:370–381
155. Shahinian VB, Kuo YF, Freeman JL, Goodwin JS (2005) Risk of fracture after androgen deprivation for prostate cancer. N Engl J Med 352:154–164
156. Smith MR, Goode M, Zietman AL et al (2004) Bicalutamide monotherapy versus leuprolide monotherapy for prostate cancer: effects on bone mineral density and body composition. J Clin Oncol 22:2546–2553
157. Oefelein MG, Ricchiuti V, Conrad W, Resnick MI (2002) Skeletal fractures negatively correlate with overall survival in men with prostate cancer. J Urol 168:1005–1007
158. Smith MR, Egerdie B, Hernandez Toriz N et al (2009) Denosumab in men receiving androgen-deprivation therapy for prostate cancer. N Engl J Med 361:745–755
159. Michaelson MD, Kaufman DS, Lee H et al (2007) Randomized controlled trial of annual zoledronic acid to prevent gonadotropin-releasing hormone agonistinduced bone loss in men with prostate cancer. J Clin Oncol 25:1038–1042

Chapter 2
Advances in Radiotherapy for Prostate Cancer Treatment

Tarun K. Podder, Elisha T. Fredman, and Rodney J. Ellis

Abstract Major categories of radiotherapy (RT) for prostate cancer (CaP) treatment are: (1) external beam RT (EBRT), and (2) brachytherapy (BT). EBRT are performed using different techniques like three-dimensional conformal RT (3D-CRT), intensity modulated RT (IMRT), volumetric modulated arc therapy (VMAT), and stereotactic body radiation therapy (SBRT), stereotactic radiosurgery (SRS) and intensity modulated proton therapy (IMPT), etc., using a variety of radiation delivery machines, such as a linear accelerator (Linac), Cyberknife robotic system, Gamma knife, Tomotherapy and proton beam machine. The primary advantage of proton beam therapy is sparing of normal tissues and organ at risks (OARs) with comparable coverage of the tumor volume. MR-Linac is the latest addition in the image-guided RT. Robot-assisted brachytherapy is one of the latest technological innovations in the field. With the advancement of technology, radiation therapy for prostate cancer can be improved using high quality multimodal imaging, robot-assistance for brachytherapy as well as EBRT. This chapter presents the advances in radiation therapy for the treatment of prostate cancer.

2.1 Introduction to Radiotherapy for Prostate Cancer

Two broader categories of radiotherapy (RT) for prostate cancer treatment are: (1) external beam RT (EBRT), and (2) brachytherapy (BT). Brachytherapy is normally performed for low-intermediate stage of prostate cancer (CaP), while EBRT can be used for any stage of CaP. However, post EBRT brachytherapy boost treatment is also practiced for intermediate-high risk CaP cases. In EBRT, the radiation beams are focused on the tumor volume to deliver a prescribed dose to destroy the

T. K. Podder (✉) · E. T. Fredman · R. J. Ellis
Department of Radiation Oncology, University Hospitals Cleveland Medical Center,
Case Western Reserve University, Cleveland, OH, USA
e-mail: tarun.podder@uhhospitals.org

© Springer Nature Switzerland AG 2018
H. Schatten (ed.), *Molecular & Diagnostic Imaging in Prostate Cancer*,
Advances in Experimental Medicine and Biology 1096,
https://doi.org/10.1007/978-3-319-99286-0_2

cancerous cells. The beams are arranged around the patient in such a way that the tumor, i.e. the target volume, receives the maximum prescribed dose sparing the normal tissue and organ at risks (OARs), as much as possible. EBRT can be performed using different techniques like three-dimensional conformal RT (3D-CRT), intensity modulated RT (IMRT), volumetric modulated arc therapy (VMAT), and stereotactic body radiation therapy (SBRT), stereotactic radiosurgery (SRS) and intensity modulated proton therapy (IMPT), etc., using a variety of radiation delivery machines, such as a linear accelerator (Linac), Cyberknife robotic system, and Tomotherapy. These machines use photon beams; whereas, proton beams are used in proton (particle) therapy machine. Better dose conformality to target volume and sparing of OARs are normally achievable using IMRT/VMAT as compared to 3D-CRT. Higher dose per fraction of treatment is delivered in SBRT (1–5 fraction) and SRS (single fraction). SRS and SBRT provide the opportunity in delivering higher biologically effective dose (BED) intending for improved clinical outcomes. The major advantage of proton beam therapy is sparing of normal tissues and OARs with comparable coverage of the tumor volume. Multimodal images, such as CT, MRI, US, PET, and SPECT, are used for tumor localization and target delineation, dosimetric planning and radiation treatment delivery. Dose computation for all EBRT plans are routinely perform on CT images. However, researchers are working on developing pseudo-CT from MRI for dose computation. MRI, PET and SPECT are important imaging modalities for delineating tumor volume. Whereas, CT (CT on the rail or cone bean CT) and MRI (MR-Linac) can be very useful in delivery of image-guided radiation therapy (IGRT). MR-Linac is the latest addition in the image-guided RT. So far, maximum magnetic strength is being tested for using MRI during delivery of RT is 1.5 T. The MR image quality is expected to be good enough for localizing tumor and OARs confidently and accurately. In brachytherapy procedure, radioactive isotopes are either permanently or temporarily placed in or at a very close proximity of the tumor volume, i.e. it works like in-side-out. For prostate cancer treatment with brachytherapy, the radioactive seeds or radiation source are placed throughout the gland using long hollow needles (about 200 mm long and 1.27–1.47 mm in diameter) to deliver prescription dose uniformly sparing OARs, i.e. rectum, bladder neurovascular bundle, and urethra. In low-dose-rate (LDR) brachytherapy, about 50–100 radioactive seeds are permanently delivered in the prostate gland. These seeds remain in the patient for the rest of the patient's life. On the other hand, radioactive sources are temporarily (for about 5–10 min) placed in the prostate gland for high-dose-rate (HDR) brachytherapy treatment. Sometimes the brachytherapy may be used as boost following the EBRT. In brachytherapy, the needle insertion and seed deposition are performed under the guidance of imaging such as transrectal ultrasonography (TRUS), fluoroscopy, CT or MRI. So far, TRUS is the most common modalities of image-guidance for LDR or HRD brachytherapy.

2.2 Advances in External Beam Radiation Therapy (EBRT) for CaP

2.2.1 EBRT with Photon Beam

For EBRT, the National Comprehensive Cancer Network (NCCN) Guidelines in Oncology recommends doses between 75.6 and 79 Gy for individuals with low-risk prostate cancer. Whereas for patients with intermediate- or high-risk disease doses up to 81 Gy are appropriate [1]. High energy (6–18MV) photon beams generated by a linear accelerator are normally used for treating CaP. 3D-CRT and IMRT/VMAT are common techniques for EBRT. Cone-beam CT (CBCT) is used for daily verification for accurate positioning of the patient. Latest developments in the EBRT are mainly improvement in target/tumor identification and delineation, accuracy in patient positioning, and accuracy of dose delivery.

Data from various studies suggest a low fractionation sensitivity of prostate cancer [2, 3] indicating an advantage might be expected from hypo-fractionated treatment regimens, either with regard to increased local control, reduced late side effects or both. In this regard, stereotactic Body Radiation Therapy (SBRT) for CaP is gaining popularity due to shortened course of treatment with similar clinical outcome as compared to IMRT/VMAT.

Early foundational experiences of moderately hypo-fractionated EBRT for low- and intermediate-risk patients. Began to emerge in the early-1990's, using doses ranging between 4.5–6 Gy per fraction, with the goal of shortening treatment time, decreasing acute toxicity and achieving equivalent tumor control [4]. Subsequent and more robust trials at the start of twenty-first century favored a more modest hypo-fractionation approach for fear of worsening severe acute toxicity, and generally demonstrated similar or slightly worse acute toxicity and comparable biochemical control [5–7]. Importantly, these trials were still performed in an earlier era in which treatments were delivered to more generous pelvic fields with a 2- or 3-dimensional conformal technique, and overall lower total doses were prescribed. Once it was demonstrated in numerous trials that dose-escalated prostate radiation resulted in improved biochemical free survival [8–11], a series of modern moderate hypo-fractionated EBRT trials have been reported with strong outcomes. Three of these four large prospective trials directly showed that hypo-fractionation was non-inferior to standard 2 Gy dosing, while the fourth failed to show a superiority (and though inferences have been suggested for non-inferiority, this does not reflect the intended trial design) [12–14]. Based on these trials, a paradigm shift of considering these regimens the new standard of care has begun. Of note, this has all been within the context of low and intermediate-risk patients. At this time, there is less data regarding the role of hypo-fractionated radiation for high risk patients, though a number of randomized control trials have begun to demonstrate similar benefits as were seen in the lower risk cohorts [15, 16].

SBRT can be thought of as a form of "extreme" hypo-fractionation, which requires a high degree of technical and clinical proficiency in order to deliver high tumoricidal doses in a conformal, stereotactic manner. Typically delivered in five fractions given every-other day, SBRT utilizes continuously modulated circular arcs of radiation to both deliver ablative doses to the desired treatment volume, as well as carve around the surrounding normal structures, such as the rectum, bladder, femoral heads and penile bulb. Single arm phase I and II studies established the feasibility and safety of delivering approximately 7 Gy per fraction for five fractions, yielding excellent biochemical control at relatively brief follow up time points (5 years) and low rates of toxicity [17]. Some have investigated treating to even higher fractional doses to achieve a higher total equivalent dose with similar safety and clinical outcomes [18, 19]. Perhaps the largest series was presented from Stanford, a phase II pooled consortium from eight centers, comprising over 1100 patients. They demonstrated that five fractions of 7–7.25 Gy per fraction delivered either every-other day or daily was highly effective for low-, intermediate and even potentially high-risk patients, with only a modest increase in acute toxicity that generally resolved within a six month follow up period. All of these individual studies laid the foundation for an ongoing national clinical trial, RTOG 0938, which was recently presented in abstract form, showing very promising safety and tolerability outcomes [20].

While both moderate hypo-fractionation and SBRT appear to be safe and effective modes of radiotherapy delivery for patients with low- and intermediate-risk prostate adenocarcinoma, no direct head-to-head clinical trial of radiation schedules has been performed. To this end, the NRG recently activated the phase III clinical trial GU-005 to address this question. Patients will be randomized to receive either hypo-fractionated IMRT with 70 Gy in 28 fractions (2.5 Gy per fraction) vs. SBRT with 36.25 Gy in five fractions (7.25 Gy per fraction) [21]. The primary endpoints of this trial include demonstrating superiority of SBRT over hypo-fractionated IMRT in both disease-free survival and acute toxicity/quality of life measures. This trial is currently open to accrual.

Some institutions have begun to lay the groundwork for the possibility of delivering ablative SBRT limited to a partial volume of the prostate. By sparing the entirety of the prostate gland from receiving high doses of radiation, as well as by potentially better sparing surrounding structures, in particular bladder and rectum, partial prostate SBRT is predicted to substantially decrease the risk of toxicity. A number of challenges however remain, and are the topics are current investigation. These include optimal image-based identification of tumor within the gland, concordance (or discordance) of image findings and histopathology and the definition of the proper partial gland target volume. It remains unknown whether any visualized suspicious regions within the prostate require definitive treatment, or whether, in the setting of a "dominant nodule" seen on MRI, any remaining small volume tumor deposits can be followed over time. Currently, a pilot phase I study is underway at University Hospitals Cleveland Medical Center in which low risk patients with a dominant nodule identified on multiparametric-MRI undergo navigational targeted and whole-gland biopsies to confirm the true presence of localized disease, and if eligible, go on to partial gland focal SBRT (Fig. 2.1).

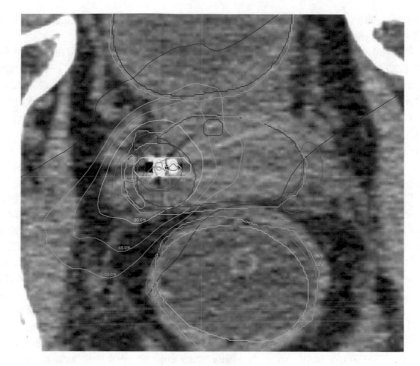

Fig. 2.1 SBRT of partial prostate using VMAT plan

There are some initiatives regarding moving forward 3D RT to 4D-RT for CaP treatment, i.e. target tracking and dynamic delivery of radiation therapy. In this regard various methods of tumor tracking, such as robotic linac (Cyberknife), robotic couch, etc. are useful [22–24]. More details about some of these techniques are provided in the later section entitled Robotic-assistance in EBRT for CaP treatment.

2.2.2   EBRT with Proton Beam

The relative biological effectiveness (RBE) is the ratio of the photon dose to the particle dose required to produce the same biological effect. For proton beams, the relative biological effectiveness is 1.1. To measure the total radiation dose with proton beams, Gray equivalents (GyE) or cobalt Gray equivalents (CGE) are often used with protons. The CGE is the Gray multiplied by the RBE factor specific for the beam used. Commonly, in the range of 78–82 CGE is prescribed for CaP treatment with proton beams, though hypofractionation may also be applied in this setting. Major advantages of proton beam therapy are expected to be higher dose deposition in tumor located in Bragg Peak region of the beam while very sharp dose falloff beyond the tumor, i.e. very low or no exit dose. This physics property of the proton beam translates to clinical advantages of better sparing of critical anatomical structures and normal tissues while dose to the tumor is not compromised.

Therefore, reduced toxicity and improved quality of life are expected when the CaP patients are treated with proton beam.

The goal of proton beam therapy is to prolong overall and disease specific survival, to delay progression of disease and to improve quality of life. Because proton beams can theoretically be more concentrated on the tumor and less concentrated on the surrounding normal tissues, theoretically it could be a treatment option for patients with localized prostate cancer that maximizes the effect on the tissue while limiting effects on adjacent tissue. The drive to further diminish toxicity and improve patient's quality of life is precisely the result of the already excellent outcomes in biochemical progression free survival and overall survival in patients with early and intermediate risk prostate adenocarcinoma.

The very first trials investigating the benefits of proton beam radiation for prostate cancer took advantage of the normal tissue-sparing properties of protons as a way to safely deliver a boost of dose to, what was at that time, a higher overall prescription than had previously been utilized [25, 26]. These trials, however, were not intended to primarily assess the role of proton radiation, rather they were studies of dose escalation, utilizing protons to facilitate the safe delivery of high doses. More recently, numerous data have emerged regarding proton beam prostate radiation, led largely by the efforts of Loma Linda University Medical Center, University of Florida, and Mass General Hospital. The majority of data shows agreement in the ability of proton radiation to effectively deliver equally high doses of radiation to the prostate compared to IMRT, produce equivalent or even moderately improved acute patient-reported toxicity, though spare substantial portions of the bladder, rectum, and surrounding pelvic tissue from the low- and intermediate integral dose bath created by multi-beam or VMAT [27–30]. The primary concern, at this time, of advocating for the application of proton beam radiation for prostate cancer is the substantially higher cost that protons carry compared to 3D-conformal radiation or IMRT. Many remain skeptical that the yet unproven and potentially limited clinical benefits outweigh this increase. Two ongoing large clinical trials, PROTECT and PARTIQoL, will hopefully shed additional light on the extent of the quality of life benefit to patients treated with protons for prostate cancer, and allow better quantification for this cost-benefit analysis [31, 32].

The design of proton beam radiation fields has evolved since the first applications in prostate cancer. At first, a single perineal field was delivered which was felt to minimize the amount of normal tissue in the beam trajectory as it travelled to the target [25]. This single field proved to be more toxic to the bladder and rectum, however, and with improvements in technology of proton delivery as well as imaging and patient setup, other permutations were evaluated. The most common angle design at this time is two opposed equally weighted lateral beams, each of which cover the entire target and range out on the respective distal side of the prostate volume (Fig. 2.2). Some attempts were made to decrease dose to the femoral heads by rotating the two beams anteriorly to create a pair of oblique entry points, and while the goal of decreasing dose to the femoral heads was accomplished, the result was an increased dose to the bladder [28]. As such, opposed laterals continue to be the orientation of choice.

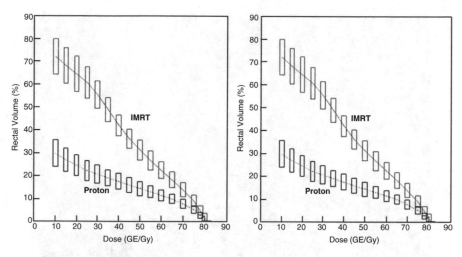

Fig. 2.2 Sparing of OARs for IMRT vs. Proton Beam plan for prostate cancer treatment [28]

2.2.3 Robotic-Assistance in EBRT for Prostate Cancer Treatment

It is commonly accepted that translation of 5 mm or more during a single treatment session is likely. Real-time tracking of the prostate has the potential to markedly improve dose delivery to tumor tissue and minimize the exposure of surrounding non-involved structures. Application of robotic systems in EBRT can be divided into three categories: (1) patient positioning, (2) radiation beam deliver, and (3) tumor tracking.

Patient Positioning: Generally, a 6 degree-of-freedom (DOF) robot with a flat table-top installed at the end-effector is used for the patient to lay down for accurate positioning under the treatment beam. A majority of cases the patient is held horizontally and depending on the tumor location and relative motion, the table-top or couch can be tilted and/or rolled to a certain angle (<5°). A 6 DOF orthomorphic arm type robotic couch is commonly used with Cyberknife and proton therapy machines. Whereas, a 6 DOF HexaPOD robotic couch is mainly used with conventional Linacs. The HexaPOD system is basically a Stewart Platform mounted on regular/standard treatment table for providing finer adjustment of patient positioning during treatment.

Radiation Beam Delivery: In this type of applications, a compact radiation source is mounted at the end-effector of the robot (Fig. 2.3). The radiation source can be a linear accelerator (Linac) that can generate photon beam for treatment. Then the compact Linac is moved around the patient to deliver treatment dose as per plan. One of such machines is the Cyberknife, which has a 6 MV X-band Linac installed at the end-effector of a 6 DOF KUKA robot [33, 34]. Cyberknife uses a pair of orthogonal X-ray imaging system for tracking a moving tumor and delivering radiation

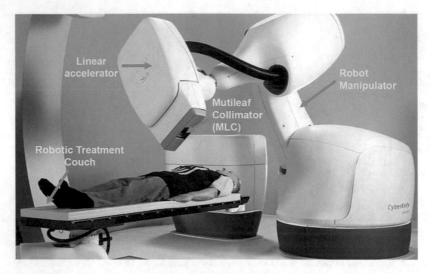

Fig. 2.3 Cyberknife robotic system for radiation therapy

dose to target tumor volume. Generally, SBRT and SRS type of treatments are done with a Cyberknife.

Tumor Tracking: One of the major challenges faced during EBRT is treating moving tumor in thoracic, abdominal, and pelvic areas. Tumors in these regions continuously move due to respiratory motion, cardiac cycle, and peristalsis of colon or rectum. There can be three types of solutions for irradiating a moving tumor: (1) use a large margin around the tumor volume for safety, so that no part of the tumor is missed during radiation dose delivery, (2) actively track the tumor by moving the radiation beam/source, (3) move the patient in opposite direction of the tumor motion, so that the tumor appears stationary with respect to the radiation beam/source. Each of these techniques has relative advantages and disadvantages over the other.

Motion of the tumors and normal tissues in thoracic, abdominal and pelvic regions due to physiological movements during radiation therapy is ubiquitous. In radiation dosimetric planning, large margins to tumors are commonly used to ensure adequate dosimetric coverage of the tumor excursion, at the expense of irradiating very large volumes of normal tissues and adjacent critical structures. Nowadays, margins to tumor are more critical because of improved survival and development of long-term side effects as well as preserving functionality or minimizing dose to critical organs for keeping the option open for re-treatment (if required) in the future. Deployment of 4D real-time tumor tracking during radiation therapy may allow smaller (or no) margin to the tumor which, in turn, can potentially reduce toxicities and improve treatment outcome as well as may permit for dose escalation in an attempt for curative treatment.

The only commercially available tracking system for radiation treatment is the Cyberknife robotic system which can be used to treat only selective patients (i.e. patients with small tumors) due to limited field sizes. Additionally, this system is not commonly available or used (less than 500 worldwide as compared to more than 25,000 linear accelerators (Linacs)), as well as patient throughput is very low, at only about 4–5 per day as compared to 25–30 par day for a Linac. Moreover, longer treatment time on Cyberknife leads to patient discomfort and more gross body movement resulting in reposition of the patient. The HexaPOD robotic couch with its 6 DOF movement can track the tumor in 3D/4D. However, it has several critical limitations: (1) small range of motion, (2) low speed, (3) mechanical structure is too tall for clinical use (it sits upon a regular couch), and (4) expensive. For these reasons, HexaPOD has not been deployed in the RT clinic for tracking. Therefore, researchers are developing robotic couch very similar to regular table used for patient positioning [23, 35]. Although robotic tracking is an attractive technique for dealing with the moving tumors, its application in prostate cancer treatment is limited. The reasons are small translational and rotational motions of the prostate gland due to patient's physiodynamics. Inter-fractional positional variations of the prostate gland with respect to other anatomical structures are more critical than intra-fractional motions. Therefore, a baseline shift is more desirable than a continuous tracking of the tumor during EBRT. However, use of electromagnetic (EM) beacons (e.g., Calypso Medical's EM sensor) along with a robotic couch or Cyberknife type robotic Linac can be useful for more accurate delivery of radiation dose with minimal margin to the tumor volume [36]. This can enable additional sparing of OARs and normal tissues which can result in improve clinical outcome including reduced toxicities, enhanced quality of life and longer survival.

2.2.4   Challenges and Potential Future Directions in EBRT

Prostate or tissue deformation and needle deflection are two major issues in accurate targeting and collection of core in prostate biopsy. Therefore, precise collection of biopsy cores from the suspected lesion and correct labeling of cores for accurate pathological reporting may be difficult. This may result in misleading diagnosis and ineffective focal therapies. This deficiency can jeopardize the implementation of focal SBRT in which the whole gland in not treated with EBRT.

With the improvement of multimodal imaging and use of multi-parametric MRI and other molecular imaging along with histopathological confirmation, the robot-assisted biopsy can potentially have high sensitivity and specificity. This can minimize the false positive or false negative results of biopsy. Finding exact location of tumor foci is very critical for focal therapy, i.e. focal SBRT or brachytherapy. However, with the advancement in imaging science, various modality of imaging such as multi-parametric MRI with high magnetic strength (above 3 T), molecular imaging, etc. may be able to replace the necessity of biopsy in the future. However due to current limitations, it may be advantageous to use a combined dataset of both anatomic and functional imaging, along with correlated histopathologic tracking from tumor biopsies to develop planning target volume

(PTV's) as is being currently validated by clinical trial for focal prostate SBRT noted previously from our institution.

There are several challenges in regards to real-time tumor tracking which can be improved by having faster imaging or sensory feedback of the spatial location of the tumor and better correlation model for the internal tumor location and external fiducial markers. Most challenging aspect in tumor tracking is the determination of tumor trajectory accurately and well in advance so that the radiation beam can be focused accurately on the target. Correlation between the external surrogates (or fiducial markers) and the internal tumor motion is always difficult. So far, the Cyberknife type robotic systems are concerned, installing a Linac with higher photon energy (>6 MV) will be challenging due to the larger size and weight associated with higher energies. Higher energies are desirable for having optimal dose distribution for prostate cancer cases, especially for bulky patients. Efficacies of 1.5 T MRI guided EBRT system in clinical environment is yet to be evaluated. Proton beam therapy for CaP is promising. However, a well-designed randomized clinical study for evaluating its clinical effectiveness in comparison to other techniques of radiotherapy such as IMRT/VAMT and brachytherapy is overdue.

2.3 Advances in Brachytherapy for CaP

Brachytherapy is delivered for CaP when the cancer is confined to the prostate. If the CaP extends beyond the prostatic capsule, into the seminal vesicles or the adjacent nodes, radical prostatectomy is sometimes combined with adjuvant radiation therapy. The main advantage of brachytherapy is the radiation dose confinement, i.e. higher sparing of normal tissues due to placement of radiation sources inside the prostate or tumor.

2.3.1 Low-Dose-Rate (LDR) Brachytherapy

In LDR brachytherapy radioactive seeds are implanted in prostate and they remain in prostate permanently. The isotope used for LDR brachytherapy are I-125, Pd-103 and Cs-131. Depending on the physical properties of the isotope the prescribed dose varies for the same biologically effective dose (BED). For example, 145, 125, 120 Gy for monotherapy and 110, 100 and 95 Gy for brachytherapy boost following EBRT for I-125 (half-life = 60 days, energy = 28 keV), Pd-103 (half-life = 17 days, energy = 21 keV) and Cs-131 (half-life = 10 days, energy = 30 keV), respectively. The dose prescription may slightly vary in different clinical practices. From clinical experience it is revealed that application of Pd-103 can reduce the treatment related toxicities as compared to I-125 isotope. Shorter half-life of the isotope speeds up the dose delivery, for example, Cs-131 can deliver 90% dose in 33 days and compared to 204 days for I-125. Therefore, shorter half-life may be advantageous for more aggressive CaP cases.

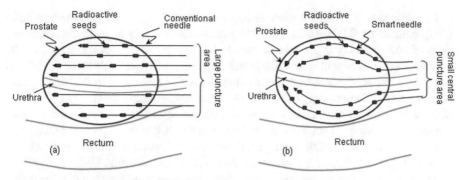

Fig. 2.4 Accessing various part of the prostate: (**a**) conventional rectilinear approach of prostate brachytherapy needle insertion pattern with straight needles requiring <u>seven needles</u> (note that the patient must be set up in the OR in the lithotomy position), and (**b**) proposed curvilinear conformal smart needle insertion requiring <u>four needles</u> [37]

The efficacy of CaP treatment with LDR brachytherapy using radioactive seeds depends on several procedural factors such as image (TRUS, MRI or CT) quality, surgical skill of the clinicians, placement of needle at desired position, deposition of the seed/source at planned location, prevention seed movement or migration.

A new technique of seed implantation has been depicted in Fig. 2.4. Here, with the new concept of curvilinear approach (Fig. 2.4b), a significant reduction of radiation dose can be achieved as compared to conventional rectilinear approach (Fig. 2.4a). With the same/similar coverage of the tumor target, an average D30 (dose to 30% volume) was reduced by 10% and D10 (dose to 10% volume) was reduced by 9.4% for urethra. In case of rectum, average reduction in V100 (volume covered by 100% of the prescription dose) was 71.6% and reduction in D5 was 17.7%. Average reductions in required number of needles and seeds (or activity) were 30.1% and 10.6%, respectively in curvilinear approach [37]. Large reduction in rectal dose would potentially reduce rectal toxicity and complications. Reduction in number of needles would minimize edema (swellings) and thereby would improve accuracy of seed delivery and total dose distribution [37] as well as reduce toxicities. Overall, this study indicated that clinical implementation of the proposed smart needle could potentially improve radiation dose distribution and reducing dose to critical organs and thereby would potentially improve quality of life and survival of the prostate cancer patients.

2.3.2   High-Dose-Rate (HDR) Brachytherapy

HDR brachytherapy may have clinical advantages over LDR brachytherapy, and it may be equally offered to patients of all risk groups. The use of HDR brachytherapy as a boost to EBRT can potentially have a lower acute toxicity than the LDR boost. This is due to the brief single fraction exposure that may be used clinically similar to that described for clinical trials or intermediate risk and high intermediate risk group patients on RTOG 0815 and RTOG 0924, respectively at 21 Gy in two fractions of 10.5 or 15 Gy in a single fraction following 45 Gy of EBRT. Data for HDR monotherapy in low- and intermediate-risk patients is also available though without supporting Phase III randomized clinical trials. A Canadian trial in this setting has been successful in accruing data on 19 Gy in a single fraction versus 27 Gy in two fractions of 13.5 Gy each though the two fraction arm was initially reported to have higher acute toxicity it remains to be determined if it has an improved bDFS rate [38]. Data comparing HDR versus LDR has shown favorable toxicity profiles for HDR over LDR monotherapy in low or intermediate risk patients [39].

HDR prostate brachytherapy is generally preformed under spinal or general anesthesia using transrectal ultrasound (TRUS) guidance. Tranperineally placement of 14–18 needles are commonly positioned with a peripheral modified distribution to avoid excess uretheral dose. Urethral visualization for placement and planning is critical to limit high urethral dose that may increase toxicity including dysuria or obstruction acutely of late stricture formation especially in the distal urethra or below the prostatic apex. In general, it is best utilized for glands less than 60 cc and low urinary retention risk (i.e. AUA/IPPS score ≤15). Once catheters are positions CT scan is typically used for reconstruction of the implant and treatment planning. Alternatively, HDR ultrasound guided placement and treatment planning may be used as developed by Elekta as the Spot system. This may provide an ability to decrease total procedure time, and theoretically could be combined with MRI/US fusion in planning to facilitate dose escalation to high grade PIRADS 4 or 5 lesions. Flouroscopic imaging of the Foley with dilute contrast (2 and 5 cc saline) can facilitate adequate placement of brachytherapy catheters to the base in addition to the sagittal ultrasound view, and or retroverted cystoscopid bladder examination to visualize tenting of the bladder mucosa against the needle tips without bladder penetration into the mucosa.

Robot-assisted Brachytherapy (LDR, HDR): In brachytherapy, imaging and dosimetric planning are vital for the delivery of high dose ionizing radiation to the prostate while minimizing the radiation exposure to adjacent organs and structures. Different imaging techniques such as TRUS, X-ray, CT, and MRI are used at different phases of the treatment. In commonly used TRUS guided brachytherapy, the dose planning can be accomplished by preplanned or intraoperative planned techniques. The preplanned technique, which is generally created a few days or weeks before the implant procedure for LDR, has limitations that may be overcome by intraoperative planning. Intraoperative planning is important to compute the correct dose to the target, mainly because preplanning is severely affected due to change in patient's position during actual brachytherapy. US, CT or MR images are used for preplanning; whereas, US images are commonly used for intraoperative planning. CT or MRI is used for post implant dosimetric evaluations. During the brachytherapy procedure, the dynamic changes also affect the accuracy of seed delivery, thereby the radiation exposure. Deformation and movement of organs or tissues are also responsible for inaccurate dose delivery. Hence dynamic intraoperative planning is important for the treatment improvement. Therefore, future advances are expected in the development of better source delivery systems, methods of enhancing seed identification, imaging techniques, and seed immobilization techniques.

In brachytherapy, positional accuracy of the radioactive seeds (or dwell position of radiation source in HDR) is very important for optimizing the dose delivery to the targeted tissues sparing the critical organs and normal tissues. But accurate steering and placement of surgical needles in soft tissues are challenging because of several reasons. Some of them are: tissue heterogeneity and elastic stiffness, unfavorable anatomy, needle bending, inadequate sensing, tissue/organ deformation and movement, poor maneuverability, and change in dynamics of various organs. In currently practiced procedures, the fixed grid holes in the template allow the surgeon to insert the needle at specified fixed positions. Very little can be done to steer the needle to a place other than straight passing through the hole in the template. Change in needle insertion position may be required based on intraoperative dynamic planning. Sometimes, especially for larger prostates, the needle needs to be angulated to avoid pubic arch interference and get access to the desired target position in the prostate for seed delivery. In current brachytherapy procedure with fixed template the needle angulation is very difficult. However, the surgeon can use a hook to bend the needle and get to the desired position after several trials. Whereas a robotic system (Fig. 2.5), with sufficient DOFs, can provide flexibility in positioning and orientating (angulating) along with improved accuracy of needle insertion and seed deposition [41]. Robot assisted therapeutic delivery system is attractive for several reasons. The main advantages are increased accuracy, reduced human variability, and possible reduction of operation time. Additionally, with the assistance of robotics system, less skillful surgeons will be able to treat patients with higher quality. The robotic systems designed and developed for prostate brachytherapy are expected to satisfy numerous objectives and functional requirements, such as (1) improve accuracy of needle placement and seed delivery, (2) improve consistency of seed implant, (3) reduce the learning curve, (4) clinician fatigue, and radiation exposure, (5) improve safety for the patient, clinicians, and the operating room (OR) staff and equipment, etc.

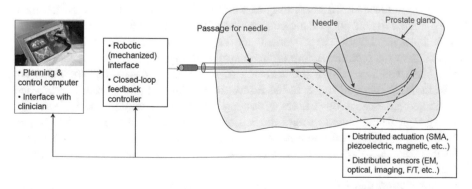

Fig. 2.5 Robot assisted rectilinear approach for LDR brachytherapy [40]

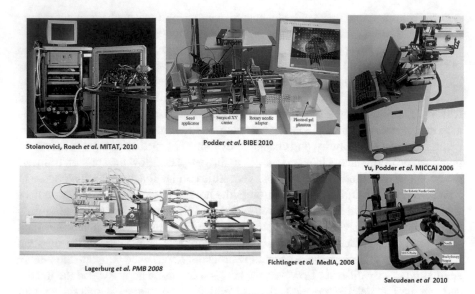

Stoianovici, Roach *et al.* MITAT, 2010

Podder *et al.* BIBE 2010

Yu, Podder *et al.* MICCAI 2006

Lagerburg *et al.* PMB 2008

Fichtinger *et al.* MedIA, 2008

Salcudean *et al* 2010

Fig. 2.6 Some of the robotic systems developed for biopsy and brachytherapy [41]

The available workspace for the robotic systems for brachytherapy is limited while the patient is in the lithotomy position for transperineal prostate brachytherapy [41]. Therefore, the robotic system should be compact, so that the clinicians' working environment is not affected. Successful clinical implementation of a robotic systems critically depends upon shape and size of the robot. Some of the brachytherapy robotic system are shown in Fig. 2.6 below.

2.3.3 Challenges and Potential Future Directions in Brachytherapy

Various sequences of MRI techniques, such as T2W, DWI, DCE and MRSI have the potential in detecting suspicious regions with higher specificity. However, these techniques need further improvement so that they can be used for definitive determination of the type and nature of the lesion. Although mpMRI is capable of identify low suspicion lesions, the accuracy of these image-based identifications are still in developmental stage. In the future, it may be possible to use image-based verification for cancerous tissues bypassing biopsy and histopathology. Therefore, using mpMRI findings to target the regions that have the highest probability of being cancer have great potential for focal therapies and radiation dose escalation. In this respect, robot-assisted MR/TRUS-guided targeted mapping biopsy can be very useful.

Application of robotic systems for prostate brachytherapy is lacking interests and supports for a variety of reasons. Accurate needle placement at a desired target in prostate is challenging due to a combining effect of tissue heterogeneity, deformation, displacement as well as needle deflection and lack of real-time accurate and faster sensory feedback as well as having better design of needles. There are significant scopes in these areas for further improvements.

2.4 Discussion/Conclusion

Both EBRT and brachytherapy are effective modalities in treating prostate cancer. In some cases, sparing of OARs seems better with proton beam therapy. With the advancement of imaging the identification of prostate lesion is becoming easier and reliable. Therefore, some studies are looking into the possibility of treating suspicious portion of the prostate rather treating the whole glad. This will potentially reduce the irradiation of normal tissue and critical structures which, in turn, is expected to reduce toxicities and improve quality of life. Both LDR and HDR brachytherapy are efficacious in treating CaP when the cancer is confined within the prostate capsule. Hypofractionated RT or SBRT is becoming popular due to shortened treatment duration without compromising clinical outcomes. The advanced technologies such as MR-Linac, robotic system, 7 T-MRI, molecular imaging, etc. have the potentially to improve the quality of radiotherapy for prostate cancer and clinical outcomes.

References

1. Kalbasi A, Li J, Berman AT et al (2015) Dose-escalated irradiation and overall survival in men with nonmetastatic prostate cancer. JAMA Oncol 1(7):897–906
2. Brenner DJ, Hall EJ (1999) Fractionation and protraction for radiotherapy of prostate carcinoma. Int J Rad Oncol Biol Phys 43(5):1095–1101
3. King CR, Fowler JF (2001) A simple analytic derivation suggests that prostate cancer a/b ratio is low. Int J Radiat Oncol Biol Phys 51(1):213–214
4. Lloyd-Davies RW, Collins CD, Swan AV (1990) Carcinoma of prostate treated byradical external beam radiotherapy using hypofractionation. Twenty-two years' experience (1962–1984). Urology 36:107–111
5. Lukka H, Hayter C, Julian JA, Warde P, Morris WJ, Gospodarowicz M, Levine M, Sathya J, Choo R, Prichard H, Brundage M, Kwan W (2005) Randomized trial comparing two fractionation schedules for patients with localized prostate cancer. J Clin Oncol 23(25):6132–6138
6. Yeoh EK, Bartholomeusz DL, Holloway RH, Fraser RJ, Botten R, Di Matteo A, Moore JW, Schoeman MN (2010) Disturbed colonic motility contributes to anorectal symptoms and dysfunction after radiotherapy for carcinoma of the prostate. Int J Radiat Oncol Biol Phys 78(3):773–780
7. Hoskin PJ, Motohashi K, Bownes P, Bryant L, Ostler P (2007) High dose rate brachytherapy in combination with external beam radiotherapy in the radical treatment of prostate cancer: initial results of a randomised phase three trial. Radiother Oncol 84(2):114–120
8. Dearnaley DP, Jovic G, Syndikus I, Khoo V et al (2014) Escalated-dose versus control-dose conformal radiotherapy for prostate cancer: long-term results from the MRC RT01 randomized controlled trial. Lancet Oncol 15(4):464–473

9. Kuban DA, Levy LB, Cheung MR, Lee AK et al (2011) Long-term failure patterns and survival in a randomized dose-escalation trial for prostate cancer. Who dies of disease? Int J Radiat Oncol Biol Phys 79(5):1310–1317

10. Michalski JM, Yan Y, Watkins-Bruner D, Walter B, Winter K, Galvin JM, Bahary J, Morton GC, M.B. Parliament, Sandler H (2011) Preliminary analysis of 3D-CRT vs. IMRT on the high dose arm of the RTOG 0126 prostate cancer trial: toxicity report. Int J Radiat Oncol Biol Phys 81(2):S1–S2

11. Pommier P, Chabaud S, Lagrange JL et al (2011) 70 Gy versus 80 Gy in localized pros-tate cancer: 5-year results of GETUG 06 randomized trial. Int J Radiat Oncol Biol Phys 80(4):1056–1063

12. Pugh J, Griffin C, Hall E et al (2016) Conventional versus hypofractionated high-dose intensity-modulated radiotherapy for prostate cancer: 5-year outcomes of the randomised, non-inferiority, phase 3 CHHiP trial. Lancet Oncol 17(8):1047–1060

13. Aluwini S, Pos F, Schimmel E, van Lin E, Krol S et al (2016) Hypofractionated versus conven-tionally fractionated radiotherapy for patients with prostate cancer (HYPRO): acute toxicity results from a randomised non-inferiority phase 3 trial. Lancet Oncol 16(3):274–283

14. Robert Lee W, Dignam JJ, Amin M, Bruner D, Low D, Swanson GP et al NRG oncology RTOG 0415: a randomized phase III non-inferiority study comparing two fractionation schedules in patients with low-risk prostate cancer. J Clin Oncol 34(2 Suppl):1. https://doi.org/10.1200/jco.2016.34.2_suppl.1

15. Arcangeli S, Strigari L, Gomellini S, Saracino B et al (2012) Updated results and patterns of failure in a randomized hypofractionation trial for high-risk prostate cancer. Int J Radiat Oncol Biol Phys 84(5):1172–1178

16. Pollack A, Walker G, Horwitz EM, Price R et al (2013) Randomized trial of hypofractionated external-beam radiotherapy for prostate cancer. J Clin Oncol 31(31):3860–3868

17. Loblaw A, Cheung P, D'Alimonte L, Deabreu A et al (2013) Prostate stereotactic ablative body radiotherapy using a standard linear accelerator: toxicity, biochemical, and pathological outcomes. Radiother Oncol 107(2):153–158

18. Mantz C (2014) A phase II trial of stereotactic ablative body radiotherapy for low-risk prostate cancer using a non-robotic linear accelerator and real-time target tracking: report of toxicity, quality of life, and disease control outcomes with 5-year minimum follow-up. Front Oncol 4:279

19. Boike TP, Lotan Y, Cho LC, Brindle J et al (2011) Phase I dose-escalation study of stereo-tactic body radiation therapy for low- and intermediate-risk prostate cancer. J Clin Oncol 29(15):2020–2026

20. Lukka H, Stephanie P, Bruner D, Bahary JP et al (2016) Patient-reported outcomes in NRG oncology/RTOG 0938, a randomized phase 2 study evaluating 2 ultrahypofractionated regimens (UHRs) for prostate cancer. Int J Radiat Oncol Biol Phys 94(1):2

21. Ellis RJ. NRG-GU005: phase III IGRT and SBRT versus IGRT and hypofractionated IMRT for localized intermediate risk prostate cancer. https://www.nrgoncology.org/Clinical-Trials/Protocol-Table

22. Detti B, Bonomo P, Masi L, Doro R, Cipressi S, Iermano C, Bonucci I, Franceschini D, Di Cataldo V, Di Brina L, Baki M, Simontacchi G, Meattini I, Carini M, Serni S, Nicita G, Livi L (2015) Cyberknife treatment for low and intermediate risk prostate cancer. Cancer Investig 33(5):188–192

23. Buzurovic I, Yu Y, Werner-Wasik M, Biswas T, Anne PR, Dicker AP, Podder TK (2012) Implementation and experimental results of 4D tumor tracking using robotic couch. Med Phys 39:6957–6967

24. Sweeney RA, Arnold W, Steixner E et al (2009) Compensating for tumor motion by a 6-degree-of-freedom treatment couch: is patient tolerance an issue? Int J Radiat Oncol Biol Phys 74:168–171

25. Duttenhaver JR, Shipley WU, Perrone T, Verhey LJ, Goitein M, Munzenrider JE, Prout GR, Parkhurst EC, Suit HD (1983) Protons or megavoltage X-rays as boost therapy for

patients irradiated for localized prostatic carcinoma. An early phase I/II comparison. Cancer 51(9):1599–1604

26. Zietman AL, Bae K, Slater JD, Shipley WU, Efstathiou JA, Coen JJ, Bush DA, Lunt M, Spiegel DY, Skowronski R, Jabola BR, Rossi CJ (2010) Randomized trial comparing conventional-dose with high-dose conformal radiation therapy in early-stage adenocarcinoma of the prostate: long-term results from proton radiation oncology group/american college of radiology 95–09. J Clin Oncol 28(7):1106–1111

27. Trofimov A, Nguyen PL, Coen JJ, Doppke KP et al (2007) Radiotherapy treatment of early-stage prostate cancer with IMRT and protons: a treatment planning comparison. Int J Radiat Oncol Biol Phys 69(2):444–453

28. Mendenhall NP, Li Z, Hoppe BS et al (2012) Early outcomes from three prospective trials of image-guided proton therapy for prostate cancer. Int J Radiat Oncol Biol Phys 82(1):213–221

29. Allen AM, Pawlicki T, Dong L, Fourkal E, Buyyounouski M, Cengel K, Plastaras J, Bucci MK, Yock TI, Bonilla L, Price R, Harris EE, Konski AA (2012) An evidence based review of proton beam therapy: the report of ASTRO's emerging technology committee. Radiother Oncol 103(1):8–11

30. Bryant C, Smith TL, Henderson RH et al (2016) Five-year biochemical results, toxicity, and patient-reported quality of life after delivery of dose-escalated image guided proton therapy for prostate cancer. Int J Radiat Oncol Biol Phys 95(1):422–434

31. Vapiwala N (2017) Surgery, radiation, or active surveillance? Findings from the ProtecT trial for prostate cancer. ASCO meeting 2017

32. Efstathiou J. A landmark study compares proton beam therapy with standard radiation therapy. http://www.massgeneral.org/cancer/advances/fall2013proton.aspx

33. Hoogeman M, Prévost J, Nuyttens J et al (2009) Clinical accuracy of the respiration tumor tracking system of the CyberKnife: assessment by analysis of log files. Int J Radiat Oncol Biol Phys 74:297–303

34. Dieterich S, Cavedon C et al (2011) Report of AAPM TG 135: quality assurance for robotic radiosurgery (CyberKnife). Med Phys 38:2914–2936

35. Buzurovic I, Yu Y, Podder TK (2011) Active tracking and dynamic dose delivery for robotic couch in radiation therapy. Proc IEEE Int Conf Eng Med Biol 2011:2156–2159

36. Zhu M, Bharat S, Michalski JM, Gay HA, Hou WH, Parikh PJ (2013) Adaptive radiation therapy for postprostatectomy patients using real-time electromagnetic target motion tracking during external beam radiation therapy. Int J Radiat Oncol Biol Phys 85:1038–1044

37. Podder TK, Hutapea P, Darvish K, Dicker AP, Yu Y (2012) A novel curvilinear approach for prostate seed implant. Med Phys 39(4):1887–1892

38. Morton G, Chung HT, McGuffin M, Helou J et al (2017) Prostate high dose-rate brachytherapy as monotherapy for low and intermediate risk prostate cancer: Early toxicity and quality-of life results from a randomized phase II clinical trial of one fraction of 19 Gy or two fractions of 13.5 Gy. Radiother Oncol 122(1):87–92

39. Martinez AA, Demanes J, Vargas C, Schour L, Ghilezan M, Gustafson GS (2010) High-dose-rate prostate brachytherapy: an excellent accelerated-hypofractionated treatment for favorable prostate cancer. Am J Clin Oncol 33(5):481–488

40. Maria Joseph F, Kumar M, Hutapea P, Yu Y, Dicker A, Podder T (2015) Development of self-actuating flexible needle system for surgical procedures. J Med Dev 9(2):1–2.020945. https://doi.org/10.1115/1.4030221 Paper No: MED-15-1138

41. Podder TK, Beaulieu L, Caldwell B, Cormack RA et al (2014) AAPM and GEC-ESTRO guidelines for image-guided robotic brachytherapy: report of Task Group 192. Med Phys 41:101501–101527

Chapter 3
Role of Prostate MRI in the Setting of Active Surveillance for Prostate Cancer

Samuel J. Galgano, Zachary A. Glaser, Kristin K. Porter, and Soroush Rais-Bahrami

3.1 Introduction

Over the past several decades, the incidence of prostate cancer has significantly increased [1, 2]. In 2017 alone there were an estimated 161,000 newly diagnosed cases in the United States [2, 3]. This is largely attributed to the increased detection of localized, low-risk cancer afforded by widespread prostate specific antigen (PSA)-based screening [1]. However, prostate cancer is a heterogeneous malignancy process with a wide spectrum of aggressiveness across patients afflicted with this disease. For men with favorable disease attributes, immediate intervention may not be necessary [4]. In fact, cancer-specific survival in patients with indolent prostate cancer is very high, even with conservative management [5]. Moreover, aggressive medical, surgical or radiation-based intervention does not come without considerable risk of harm to quality of life measures [6, 7]. Therefore, a less invasive management approach such as active surveillance may be more appropriate for certain individuals with more indolent forms of prostate cancer. It is because of this considerable variation that a number of methods have been developed for the risk-stratification of prostate cancer cases.

The initial concept behind risk-stratification in prostate cancer stems from the pathologic grading of the Gleason score, initially developed in the 1960s. The

S. J. Galgano · K. K. Porter
Department of Radiology, University of Alabama at Birmingham, Birmingham, AL, USA

Z. A. Glaser
Department of Urology, University of Alabama at Birmingham, Birmingham, AL, USA

S. Rais-Bahrami (✉)
Department of Radiology, University of Alabama at Birmingham, Birmingham, AL, USA

Department of Urology, University of Alabama at Birmingham, Birmingham, AL, USA
e-mail: sraisbahrami@uabmc.edu

© Springer Nature Switzerland AG 2018
H. Schatten (ed.), *Molecular & Diagnostic Imaging in Prostate Cancer*,
Advances in Experimental Medicine and Biology 1096,
https://doi.org/10.1007/978-3-319-99286-0_3

Gleason score (since revised in 2005 and 2014 by the International Society of Urologic Pathology to the modified Gleason score) describes the pathologic findings of prostate cancer on microscopy, with scores given to the two most common patterns observed ranging from 1 to 5 each [8–10]. The most recent revision in the Gleason scoring system does not allow for Gleason sums of 2–5 to be assigned to biopsy specimens, resulting in prostate cancers diagnosed at biopsy assigned Gleason scores ranging from 6 to 10 [10]. The Gleason score was the first effort made to link the histopathologic findings of prostate cancer with disease prognosis, with a higher Gleason score indicating a poorer prognosis [11, 12]. Although the Gleason score was an important first step in the risk-stratification of patients with prostate cancer, diagnosis and treatment of prostate cancer has evolved over time and the Gleason score (and modified Gleason score) are somewhat problematic in current practice. As a result, there has been development of a new pathologic grading system for the prognostication of patients with prostate cancer referred to as "Grade Groups" [13]. The new classification system reduces the number of Grade Groups to five, which are determined by their Gleason patterns and sums as found on microscopy, and demonstrates five distinct rates of 5-year biochemical recurrence-free survival following radical prostatectomy [13].

As an adjunct to prognostication by the Gleason score and Grade Group alone, a variety of risk-stratification models have been developed, clinically validated, and subsequently revised, including Partin tables, Briganti nomograms, and the Memorial Sloan Kettering Cancer Center nomogram all largely incorporating the Gleason score with other patient-specific parameters proven as prognosticators in prostate cancer risk determination [14–18]. These risk-prediction models incorporate the clinical stage of the tumor and the serum PSA level in addition to the biopsy Gleason score, which is consistently the most significant prognosticating factor, to provide population-based estimates of probability of organ-confined disease, extraprostatic extension, seminal vesicle invasion, and lymph node involvement. A recent head to head meta-analysis demonstrated no significant difference between the three models above in the prediction of lymph node involvement [19]. These risk calculators are used to help guide clinical decision-making and to assist in the determination of the likelihood of a satisfactory surgical outcome. The integration of the serum PSA into the risk-stratification is important, as prostate cancer is often multifocal within the gland and some uncertainty remains if the most aggressive lesion was sampled on a random systematic 12-core extended-sextant biopsy.

Based on the data from the risk-prediction models for patients with low-grade prostate cancer (e.g. Grade Group 1), there is a high likelihood of organ-confined disease. The concept of active surveillance was first proposed in 1992 where a cohort of men with organ-confined prostate cancer were offered conservative management instead of radical prostatectomy or external beam radiation therapy and demonstrated no difference in overall age-adjusted survival between the cohorts [20]. For patients managed with active surveillance in this early report, a serum PSA and digital rectal exam (DRE) were performed every 6 months to evaluate for disease stability. However, this initial study only followed patients for 3 years, and subsequent studies with longer follow-up demonstrated more patients with advanced disease at 10-year follow-up [21]. More recent research demonstrates uncertainty about potential mortality reduction between surgical intervention and active surveillance for select patients [22–24]. Five-year outcomes following both radical prostatectomy and external beam radiation therapy show significant morbidity due to urinary incontinence and erectile dysfunction [25]. Therefore, patients are at risk for significant morbidity if treated too aggressively, but at risk for developing advanced, potentially incurable disease if inappropriately risk-stratified and choosing to pursue active surveillance. Results of a long-term multicenter international study have validated the use of active surveillance in patients with low-risk prostate cancer to reduce overtreatment [26]. Additionally, while there is increased risk of progression to clinically-significant disease, there is an increasing body of evidence to support the use of active surveillance in some highly selected, intermediate-risk prostate cancer patients with favorable features [27–30].

3.2   Patient Selection for Active Surveillance

The underlying principle of active surveillance is to safely monitor a patient who carries the diagnosis of prostate cancer with the intention to treat the disease when necessary, while at the same time not allowing disease progression beyond the window of potential cure with definitive treatment approaches. The decision to place a patient on active surveillance versus undergoing primary definitive treatment is based on patient comorbidities, tumor characteristics, and patient/provider preferences. Life expectancy of the patient is critical in the setting of newly diagnosed prostate cancer. Those who are asymptomatic (i.e. no voiding symptoms or signs of metastatic disease) with an estimated life expectancy of less than 5 years most likely do not require intervention [31]. For men with a life expectancy of ≥10 years, assessment of tumor burden and proper risk-stratification is crucial to guiding safe patient selection for active surveillance (Table 3.1).

In 1994, Epstein and colleagues first described criteria to characterize low-risk prostate cancer in a cohort of 660 men who underwent radical prostatectomy [32]. They described 'insignificant' tumors as having a Gleason score ≤6, clinical stage T1c or less, and either a PSA density (PSAD) of <0.1 ng/mL/g without a positive

Table 3.1 Prostate cancer Grade Groups, corresponding Gleason scores, and rates of 5-year biochemical recurrence-free survival following radical prostatectomy [13]

Grade Group	Gleason score(s)	Rate of 5-year biochemical recurrence-free survival following radical prostatectomy (%)
1	3 + 3 = 6	96
2	3 + 4 = 7	88
3	4 + 3 = 7	63
4	8	48
5	9–10	26

Table 3.2 Very low-risk and low-risk prostate cancer as defined by the 2016 NCCN Guidelines, Prostate Cancer [31]

	PSA (ng/mL)	PSA density (ng/mL/g)	TNM stage	Gleason score	Biopsy cores (max no. pos.; % cancer/core)
Very low	≤10	≤0.15	≤T1c	3 + 3	3; 50
Low	≤10	≤0.15	≤T2a	3 + 3	3; 50

Table 3.3 Various institutional criteria for active surveillance patient selection [26, 36–38, 59, 100]

	Johns Hopkins University	PRIAS	UCSF	University of Toronto	Royal Marsden	Memorial Sloan Kettering
PSA (ng/mL)	≤10	≤10	≤10	≤10	≤10	≤10
PSA density (ng/mL/g)	≤0.15	≤0.20				
TNM stage	≤T2a	≤T2a	≤T2a		≤T2a	≤T2a
Gleason score	3 + 3	3 + 3	3 + 3	3 + 3	≤3 + 4	3 + 3
Biopsy cores (max no. pos.; % cancer/core)	2; 50	2; % not considered	1/3 total cores; 50		1/2 total cores; % not considered	3; 50

trans-rectal ultrasound (TRUS) biopsy core *or* PSAD < 0.15 ng/mL/g with low volume cancer and limited number core involvement [32]. In a similar time period, an additional study demonstrated that patients with low-grade tumor volumes of less than 0.5 mL were unlikely to develop clinically-significant prostate cancer in their lifetime [33]. In a similar study 4 years later, D'Amico et al. defined low-risk prostate cancer as having a PSA ≤ 10, Gleason score ≤ 6, and clinical stage ≤T2a [34]. The American Urological Association, Society of Urologic Oncology, American Society of Radiation Oncology, and National Comprehensive Cancer Network (NCCN) currently suggest active surveillance is a safe option for men who meet the criteria of very low-risk or low-risk prostate cancer designations [31, 35]. Tumor characteristics meeting very low-risk and low-risk according to the most recent NCCN guideline can be seen in Table 3.2.

Multiple institutions have introduced their own modified patient selection criteria for active surveillance (Table 3.3). For example, the Johns Hopkins University criteria only permits ≤2 positive cores with ≤50% positivity in each core, assuming all other criteria met [36]. Conversely, the University of California at San Francisco (UCSF) selection criteria includes patients who have positive prostate cancer findings on up to one third of biopsy cores sampled [37]. At the Royal Mardsen Hospital, Selvadurai et al. expanded their active surveillance criteria to include men 65 years or older with Gleason score 3 + 4 from their initial prostate biopsy [38]. Over a mean follow-up time of 5.7 years, their cohort of 471 men (only 33 of whom had

Gleason 3 + 4 disease) exhibited satisfactory progression free survival [38]. However, among those with initial Gleason 3 + 4 disease, nine (27%) were found to have worsened histology on subsequent biopsy thus bringing into question the appropriateness of broadening active surveillance criteria to include intermediate risk disease [38].

Perhaps the greatest degree of variability between clinical guidelines and practice patterns arises from patient and provider preferences following the initial diagnosis of prostate cancer. Younger men with favorable tumor characteristics may still opt for aggressive surgical or nonsurgical intervention due to factors such as a fear of living with cancer, a poor prior experience with or perception of cancer, or a failure of the provider to appropriately counsel the risks and benefits of active surveillance [6, 39]. Furthermore, men with anxiety at the time of baseline diagnosis tend to tolerate active surveillance less favorably than the general population. However, in a large prospective multi-institutional cohort of greater than 500 European men in the Prostate Cancer Research International: Active Surveillance (PRIAS) study, psychological factors did not contribute to the decision to depart from active surveillance [40]. Nonetheless, in-depth counseling at the time of offering active surveillance should be underscored as a critical element in allowing patients to make an informed decision regarding the management option they elect to pursue. Furthermore, support through available prostate cancer support groups may serve to minimize patient anxiety about their diagnosis and treatment selection [6].

3.3   Monitoring Strategies for Men on Active Surveillance

As with patient selection criteria, surveillance strategies for men on active surveillance vary across institutions. Following the initial PSA, DRE, and prostate biopsy that generated the diagnosis of prostate cancer, providers must decide when to obtain a repeat PSA, perform a subsequent biopsy and also whether or not to obtain additional diagnostic information by way of biomarkers and advanced imaging. Since tumor grade is considered the best predictor of underlying prostate cancer biology, repeated biopsies have been the hallmark of active surveillance. Retrospective analyses estimate cancer under-grading occurs an upwards of 20–30% of the time [41, 42]. For this reason, many urologists advocate for a repeat confirmatory biopsy within 3–6 months of initial diagnosis to avoid this possible oversight. This strategy has more recently been extended to an initial confirmatory biopsy being recommended by some groups within 12–24 months of initial diagnosis [43–45]. Following the initial diagnostic period, most urologists will monitor serum PSA, DRE, and repeat prostate biopsies at an interval of 1–2 years [46]. This is continued until cancer progression is detected or active surveillance is no longer necessary due to patient age dictating shortened life expectancy or other competing risks based on comorbidities affecting life expectancy. To reduce the possibility of Gleason score misclassification, extended template biopsies (i.e. transrectal or transperineal saturation) are sometimes considered prior to active surveillance enrollment, but data in support of this practice have been variable [47–50].

Recently, a plethora of genetic assays have become available to risk stratify patients with newly diagnosed low-risk prostate cancer who are considering active survcillance. For example, the 4Kscore test comprises plasma total PSA, free PSA, intact PSA and kallikrein-2 along with several clinical factors, and has been shown to reliably predict Gleason score upgrading on the first surveillance biopsy [51]. The Oncotype Dx panel is a 17-gene diagnostic assay that has been shown to reliably predict adverse pathology on radical prostatectomy specimens of men with clinically low-risk prostate cancer [52]. The Prolaris test evaluates the RNA expression of numerous genes related to cell cycle progression and other factors promoting tumor growth, and may help providers determine which patients may safely lengthen interval for repeated biopsies [53]. Evidence in support of panels such as these is limited; although, it is accumulating. Additionally, factors such as patient/provider awareness, accessibility to these advanced biomarkers, and potential out-of-pocket costs must be taken into consideration on a case-by-case basis.

Defining disease progression while on active surveillance varies across institutions. The determination has traditionally been based on discovery of higher-grade disease on biopsy, meeting or exceeding a certain PSA threshold, worrisome PSA kinetics (rate of PSA increase over a given time), and/or a change in DRE. However, finding a change in Gleason score on repeat biopsies or DRE is unusual in someone with truly very low-risk or low-risk disease [36, 46]. Instead, an upward stage migration on repeat biopsy may be due to 'missclassification' of disease severity during the initial workup and templated biopsy [54]. To make matters more challenging, using a definitive PSA threshold or cutoff in PSA kinetics has not been shown to reliably predict disease progression in several large retrospective cohorts of men on active surveillance [55, 56].

In practice, the decision to depart from active surveillance and undergo delayed intervention is either due to violation of the chosen eligibility criteria or unwillingness of the patient to continue on surveillance (an estimated 13% of cases) [57]. Currently, a plethora of evidence suggests using any of the widely accepted active surveillance enrollment criteria offers equivocal survival to immediate intervention. Using a hypothetical cohort of men eligible for active surveillance based on Johns Hopkins University criteria, Xia et al. calculated a mean survival benefit of just 1.8 months for those undergoing immediate intervention, while opting for active surveillance would confer a mean 6.4 years without invasive treatment [58]. In a cohort of 450 men on active surveillance under the University of Toronto criteria with a median follow-up of 6.8 years, Klotz et al. observed only five prostate cancer-specific deaths. Notably, this cohort included a select subset of men with intermediate-risk disease [59]. After a decade of follow-up of men on active surveillance in the PRIAS study, 52% and 73% of men had discontinued active surveillance at 5 and 10 years, respectively [60]. However, nearly one third of those men who underwent radical prostatectomy had favorable pathology (Gleason 6 disease and pT2), highlighting the need for better risk-stratification modalities [60].

3.4   Role of Magnetic Resonance Imaging for Active Surveillance

Recently, multiparametric magnetic resonance imaging (mpMRI) of the prostate has become a relied upon diagnostic and monitoring adjunct for men on active surveillance [61–63]. The Prostate Imaging Reporting and Data System (PI-RADS), and more recently PI-RADS v.2.0 and updated iterations, have become useful tools to evaluate a suspicious prostatic neoplasms [64, 65]. While data obtained from mpMRI has not yet been incorporated into existing active surveillance protocols, early evidence suggests using various PI-RADS cutoffs may increase the detection of clinically significant prostate cancer [66–69]. There is even some evidence to suggest PI-RADS v.2.0 may reliably predict active surveillance failure in certain men [70]. However, multiple studies have revealed limitations in the PI-RADS algorithm to detect all forms of significant prostate cancer, especially for men with intermediate risk (i.e. PI-RADS 3) or central zone lesions [71, 72].

Using MRI-US fusion technology to perform prostate biopsies utilizes the improved anatomical visualization afforded by this advanced imaging modality to perform targeted sampling of suspicious prostatic lesions. Several investigations have demonstrated a superior ability of MRI-US fusion compared to standard template TRUS biopsies to detect new clinically-significant prostate cancer foci in men on active surveillance [73–77]. However, it has been argued that combining standard template TRUS-guided sampling in men with MRI-targeting is still necessary, as the template cores can identify significant cancers missed by MRI-targeting alone [73–75, 78].

Determining the optimal interval for repeated imaging and possible biopsy is an ongoing process, and generally varies based on specific patient characteristics as well as institutional protocols. There is emerging evidence to suggest serial mpM-RIs alone may sufficiently monitor low-risk men on active surveillance for certain periods of time and that the interval for confirmatory biopsies may, as a result, be extended [79–84]. Based on the natural history of small index lesions of the prostate, the group at the National Cancer Institute of the National Institutes of Health suggest that forgoing any form of surveillance for up to 2 years is safe for men meeting certain mpMRI parameters [82]. However, it is important to note that these studies are largely retrospective in nature, and evaluation in the prospective setting with significant follow-up periods are necessary.

3.5   Technical Aspects of Prostate MRI for Active Surveillance

Performance of a high-quality mpMRI is becoming a cornerstone of patient management and accurate risk-stratification for active surveillance. To provide guidance and promote standardization of technique, the American College of Radiology

(ACR) published PI-RADS v.2.0, which includes a dedicated section detailing the technical parameters of mpMRI [85]. PI-RADS v.2.0 recommends that all mpMRI be performed on either a 1.5 Tesla (T) or 3 T MRI, with the 3 T MRI offering superior spatial and temporal resolution compared to the 1.5 T. However, use of a 1.5 T MRI for mpMRI may be helpful in cases of implanted devices that are MRI conditional at 1.5 T and not 3 T or in the case where artifact from adjacent implanted devices (e.g. bilateral metallic hip prostheses) may interfere with interpretation of a 3 T mpMRI.

Patient preparation for mpMRI is relatively straightforward. Patients are asked to evacuate their rectum prior to the exam to minimize artifact from air within the rectal vault. Some institutions advocate the use of an endorectal coil (ERC) to increase image signal-to-noise, but there is considerable variation in the practice patterns between institutions. The use of an ERC may potentially be most useful in the performance of mpMRI on 1.5 T machines, but comparable images can be produced on a 3 T MRI and diagnostic 1.5 T images can be obtained without the use of an ERC. Additionally, the placement of an ERC and inflation of the ERC balloon may also result in artifacts if not carefully performed. There is no strict guideline in PI-RADS v.2.0 regarding the use of an ERC; obtaining optimal images from the available MRI equipment is paramount. To reduce motion artifact from bowel peristalsis, some institutions use an antispamodic agent (scopolamine, glucagon, or hyoscyamine). However, these agents are not necessary in many patients and cost, potential benefit, and potential for adverse drug reaction should be evaluated.

Postbiopsy hemorrhage is a common finding on mpMRI, particularly if the exam is being performed to evaluate and confirm safe candidacy for active surveillance. Hemorrhage manifests as increased T1 signal at locations that have been previously sampled by systematic TRUS-guided needle biopsy and may interfere with the interpretation of mpMRI. If the mpMRI is being performed to evaluate for additional lesions within the prostate that may be a higher grade than the prostate cancer diagnosed on conventional 12-core random biopsy, the mpMRI should be performed immediately, as the hemorrhage from the biopsy should not obscure the detection of lesions potentially missed on random biopsy. However, if a biopsy-proven lesion needs staging to evaluate candidacy for active surveillance, consideration should be given to delaying the mpMRI for at least 6 weeks post biopsy to assess for resolution of the hemorrhage [86].

3.6 Interpretation of Prostate MRI in the Setting of Active Surveillance

Unlike the American College of Radiology Breast Imaging Reporting and Data System (BI-RADS), PI-RADS does not delineate a separate category or interpretation in the setting of known malignancy. Therefore, it is important for the reader to carefully examine each MRI in the setting of active surveillance for the development of new suspicious lesions. While many prostate MRIs are performed for the detection of lesions in the setting of an elevated PSA, patients on active surveillance undergo repeat MRIs and particular attention must be given to changes that develop in the time interval between scans. Similar to the initial interpretation, knowledge of PSA values and their stability is particularly helpful to the reader.

The vast majority of patients being considered for active surveillance in current clinical practice harbor Gleason 6 disease. It is well-established that prostate MRI is not sensitive for the detection of Gleason 6 cancer, but has a high negative predictive value for ruling out clinically-significant prostate cancer (Gleason ≥7) [87, 88]. As such, prostate MRI has been shown to be helpful in the detection and characterization of higher-grade, clinically-significant prostate cancer [89]. Prostate MRI, therefore, has a pivotal role in determining the eligibility of a patient for active surveillance by noninvasively excluding under-sampling of potentially clinically-significant prostate cancer [68]. Studies have also demonstrated that prostate MRI outperforms serum PSA values in the detection of prostate cancer [90].

Various studies have explored the utility and value of prostate MRI for predicting disease course, particularly for those patients with Gleason 6 prostate cancer on active surveillance [81]. Risk-prediction models and nomograms have been proposed, which integrate MRI findings with PSA values and PSA densities [69]. A major advantage to monitoring patients on active surveillance with prostate MRI is the potential decreased number of repeat biopsies needed to ensure stability, with some studies estimating the potential reduction to be up to 68% [80]. A stable prostate MRI is associated with a stable Gleason score and provides a noninvasive confirmation of disease stability to allow a patient to continue on active surveillance [84, 91].

Because many patients considered for active surveillance most commonly harbor Gleason 6 prostate cancer, their prostate MRI often does not demonstrate a highly suspicious (PI-RADS 4 or 5) lesion. Frequently, the known cancer is classified as PI-RADS 3 or intermediate suspicion for clinically significant prostate cancer. For example, in the peripheral zone of the prostate gland, these lesions may have an obvious T2-weighted signal abnormality but only mild or questionable restricted diffusion, rendering them intermediate suspicion (Fig. 3.1). Again, the major utility of prostate MRI in the initial assessment of a patient for active surveillance is the exclusion of under-sampled clinically-significant lesions which would prompt definitive treatment (Fig. 3.2). In the follow-up setting, a stable MRI appearance of the prostate would allow patients to continue on active surveillance, whereas the presence of a new suspicious lesion or suspicious change in the known prostate cancer may prompt a biopsy (Fig. 3.3). It is also important for the prostate MRI interpreter to understand the treatment that the patient is on (if any), as research shows that dutasteride results in increases in tumor ADC and decreased lesion conspicuity [92]. In addition to careful inspection of the prostate gland on MRI, close attention must be given to the pelvic lymph nodes and seminal vesicles, as they frequently are the first site of metastases in patients whose prostate cancer has

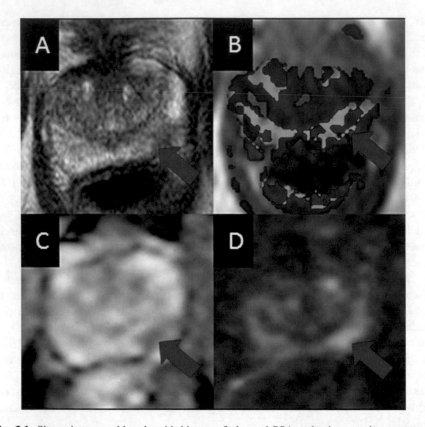

Fig. 3.1 Sixty-nine-year-old male with history of elevated PSA and prior negative systematic biopsy. The patient's most recent PSA was 9.0. mpMRI demonstrates a left posterior lateral mid gland moderately hypointense lesion on T2-weighted images (**a**, red arrow) with corresponding moderate hypointensity on ADC (**c**, red arrow) and mild hyperintensity on high b-value DWI (**d**, red arrow). This lesion did not demonstrate corresponding abnormal perfusion (**b**, red arrow) and is consistent with PI-RADS 3, intermediate suspicion for clinically significant prostate cancer. The patient underwent MRI-US fusion biopsy with this target yielding prostatic adenocarcinoma, Gleason score 3 + 3 = 6 (grade group 1), involving 2 cores (20%, 10%)

progressed (Fig. 3.4). Again, knowledge of serum PSA is extremely helpful and a significant increase in serum PSA in patients on active surveillance should prompt high suspicion for progression of disease.

3.7 Potential Role of Prostate MRI for MRI-US Fusion Biopsies

As stated previously, prostate MRI plays an essential role in the initial determination of active surveillance eligibility by ensuring that the most suspicious lesion has been adequately sampled. However, in the setting of suspicious lesion

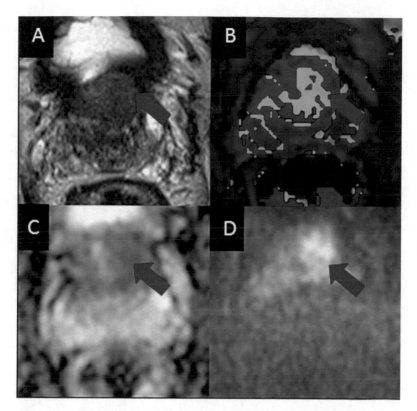

Fig. 3.2 Sixty-six-year-old male with history of elevated PSA (9.85) and systematic biopsy with Gleason 3 + 3 = 6 (grade group 1) in the left apex and left lateral apex (involving less than 5% of one core and discontinuously involving 60% of one core, respectively). mpMRI obtained to further assess suitability for active surveillance demonstrates a left base anterior transition zone lesion that is non-circumscribed, homogeneous, and moderately hypointense on T2-weighted images (**a**, red arrow). This lesion measures 1.2 cm in greatest dimension and has focal moderate hypointensity on ADC (**c**, red arrow) and mild hyperintensity on high b-value DWI (**d**, red arrow). This lesion also demonstrates asymmetric abnormal perfusion (**b**, red arrow) and is PI-RADS 4, high suspicion for clinically significant prostate cancer. The patient underwent MRI-US fusion biopsy with this target yielding prostatic adenocarcinoma, Gleason score 3 + 4 = 7 (grade group 2), involving two cores (90%, 90%)

under-sampling, MRI-US fusion-guided biopsies can be performed. Regardless of the eligibility of the patient for active surveillance, MRI-US fusion-guided biopsies have been shown to increase the detection of clinically-significant prostate cancer and decrease the detection of clinically-insignificant prostate cancer [93]. A study examining the addition of MRI-US fusion-guided biopsy to a standard 12-core biopsy demonstrated a significant increase (approximately 30%) in the number of men who were ineligible for active surveillance [94]. Additionally, MRI-US fusion-guided biopsies allow for potential electronic tracking of tumor sites and for repeated sampling of the same clonal tumor focus [95, 96]. If a patient is being

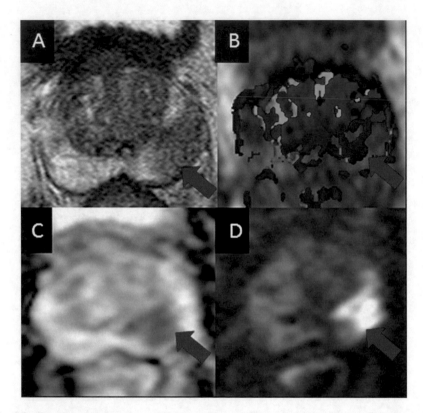

Fig. 3.3 Seventy-year-old male on active surveillance with small volume Gleason 6 disease in the right base and right mid-gland diagnosed on systematic biopsy. The patients most recent PSA was 7.3, which is up from prior 4.8, prompting mpMRI. mpMRI demonstrates a left mid gland posterior lateral peripheral zone lesion with focal, marked hypointensity on ADC (**c**, red arrow) and marked hyperintensity on high b-value DWI (**d**, red arrow). This lesion measures greater than 1.5 cm, is hypointense on T2-weighted images (**a**, red arrow) and demonstrates corresponding abnormal perfusion (**b**, red arrow). This lesion is very high suspicion for clinically significant prostate cancer (PI-RADS 5) and has extracapsular extension. The patient underwent MRI-US fusion biopsy with this target yielding prostatic adenocarcinoma, Gleason score 4 + 3 = 7 (grade group 3), involving 30% of one core and Gleason score 3 + 4 = 7 (grade group 2), involving two cores (20%, 50%). Perineural invasion was present

considered for focal treatment of their prostate cancer, MRI-US fusion-guided biopsies can help characterize the desired lesion and assist in treatment planning [86].

MRI-US fusion-guided biopsies play a significant role in the active surveillance treatment algorithm. MRI-US fusion-guided biopsies have been shown to outperform standard 12-core systematic biopsies and demonstrate a correlation between tumor volume and highest percentage of cores involved and core tumor length [97]. For patients on active surveillance, MRI-US fusion-guided biopsies almost doubled the detection of progression compared to standard 12-core systematic biopsies. Further, progression on mpMRI was the sole predictor of pathological progression in patients on active surveillance [79]. Prostate MRI demonstrates the potential to

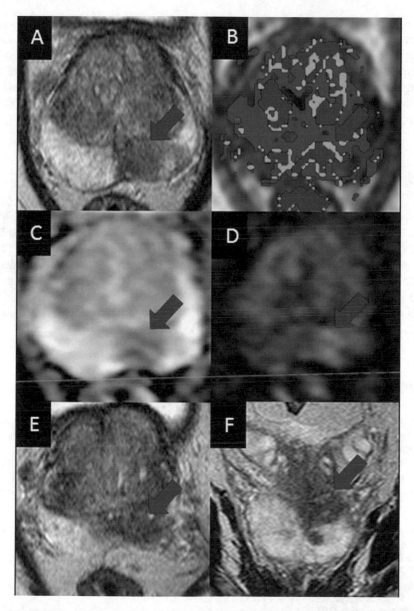

Fig. 3.4 Sixty-nine-year-old male presents for evaluation of elevated PSA of 7.3, previously 4.0. The patient is biopsy naïve. Left mid to base posterior medial peripheral zone lesion with focal marked hypointensity on ADC (**c**, red arrow) and marked hyperintensity on high b-value DWI (**d**, red arrow). This lesion measures greater than 1.5 cm and is homogeneously hypointense on T2-weighted images (**a**, red arrow) with abnormal perfusion (**b**, red arrow). This lesion is very high suspicion for clinically significant prostate cancer (PI-RADS 5) and has extracapsular extension (**e**, red arrow) with involvement of the seminal vesicles (**f**, red arrow). The patient underwent 12-core extended sextant TRUS-guided prostate biopsy and MRI/US fusion biopsy of MRI areas of suspicion. The target lesion in the left posterior medial peripheral zone base, including the seminal vesicle, yielded prostatic adenocarcinoma, Gleason score 4 + 3 = 7, involving three cores (100%, 80%, 80%). Subsequent radical prostatectomy surgical pathology showed that the bilateral seminal vesicles were positive for carcinoma

decrease the number of unnecessary biopsies for men on active surveillance and allows for more targeted sampling of the prostate gland to ensure diagnostic accuracy [84]. MRI-US fusion-guided biopsies have been shown to provide actionable information in approximately 25% of men [98]. Electronic tracking of prostate cancers during active surveillance with MRI-US fusion-guided biopsies has been shown to lead to pathologic upgrading more frequently than with non-targeted biopsies [99].

3.8 Summary

Active surveillance is a cornerstone of patient management for patients with low-risk and some intermediate-risk prostate cancers. An essential aspect of improving outcomes, morbidity, and mortality for patients on active surveillance involves careful patient selection. Multiparametric prostate MRI is a valuable tool that can be used both for initial evaluation of patients with low-risk and intermediate-risk prostate cancer, targeted biopsies, and serial imaging surveillance. Knowledge of the findings that would prompt a change in management and potential withdrawal from active surveillance is important for both urologists and radiologists.

References

1. Etzioni R, Gulati R, Cooperberg MR, Penson DM, Weiss NS, Thompson IM (2013) Limitations of basing screening policies on screening trials: The US Preventive Services Task Force and prostate cancer screening. Med Care 51(4):295–300
2. Jemal A, Ward EM, Johnson CJ et al (2017) Annual report to the nation on the status of cancer, 1975–2014, featuring survival. J Natl Cancer Inst 2017:109(9)
3. Siegel RL, Miller KD, Jemal A (2017) Cancer statistics, 2017. CA Cancer J Clin 67(1):7–30
4. Ritch CR, Graves AJ, Keegan KA et al (2015) Increasing use of observation among men at low risk for prostate cancer mortality. J Urol 193(3):801–806
5. Albertsen PC, Hanley JA, Fine J (2005) 20-Year outcomes following conservative management of clinically localized prostate cancer. JAMA 293(17):2095–2101
6. Penson DF (2012) Factors influencing patients' acceptance and adherence to active surveillance. J Natl Cancer Inst Monogr 2012(45):207–212
7. Resnick MJ, Koyama T, Fan KH et al (2013) Long-term functional outcomes after treatment for localized prostate cancer. N Engl J Med 368(5):436–445
8. Gleason DF (1966) Classification of prostatic carcinomas. Cancer Chemother Rep 50(3):125–128
9. Epstein JI, Allsbrook WC Jr, Amin MB, Egevad LL, Committee IG (2005) The 2005 International Society of Urological Pathology (ISUP) consensus conference on Gleason grading of prostatic carcinoma. Am J Surg Pathol 29(9):1228–1242
10. Epstein JI, Egevad L, Amin MB et al (2016) The 2014 International Society of Urological Pathology (ISUP) consensus conference on Gleason grading of prostatic carcinoma: definition of grading patterns and proposal for a new grading system. Am J Surg Pathol 40(2):244–252
11. Mellinger GT, Gleason D, Bailar J 3rd. (1967) The histology and prognosis of prostatic cancer. J Urol 97(2):331–337

12. Gleason DF, Mellinger GT (1974) Prediction of prognosis for prostatic adenocarcinoma by combined histological grading and clinical staging. J Urol 111(1):58–64
13. Epstein JI, Zelefsky MJ, Sjoberg DD et al (2016) A contemporary prostate cancer grading system: a validated alternative to the Gleason score. Eur Urol 69(3):428–435
14. Huang Y, Isharwal S, Haese A et al (2011) Prediction of patient-specific risk and percentile cohort risk of pathological stage outcome using continuous prostate-specific antigen measurement, clinical stage and biopsy Gleason score. BJU Int 107(10):1562–1569
15. Makarov DV, Trock BJ, Humphreys EB et al (2007) Updated nomogram to predict pathologic stage of prostate cancer given prostate-specific antigen level, clinical stage, and biopsy Gleason score (Partin tables) based on cases from 2000 to 2005. Urology 69(6):1095–1101
16. Briganti A, Chun FK, Salonia A et al (2006) Validation of a nomogram predicting the probability of lymph node invasion among patients undergoing radical prostatectomy and an extended pelvic lymphadenectomy. Eur Urol 49(6):1019–1026 discussion 1026–1017
17. Briganti A, Larcher A, Abdollah F et al (2012) Updated nomogram predicting lymph node invasion in patients with prostate cancer undergoing extended pelvic lymph node dissection: the essential importance of percentage of positive cores. Eur Urol 61(3):480–487
18. Cagiannos I, Karakiewicz P, Eastham JA et al (2003) A preoperative nomogram identifying decreased risk of positive pelvic lymph nodes in patients with prostate cancer. J Urol 170(5):1798–1803
19. Cimino S, Reale G, Castelli T et al (2017) Comparison between Briganti, Partin and MSKCC tools in predicting positive lymph nodes in prostate cancer: a systematic review and meta-analysis. Scand J Urol 51(5):345–350
20. Jones GW (1992) Prospective, conservative management of localized prostate cancer. Cancer 70(1 Suppl):307–310
21. Gerber GS (1994) Conservative approach to the management of prostate cancer. A critical review. Eur Urol 26(4):271–275
22. Bill-Axelson A, Holmberg L, Garmo H et al (2014) Radical prostatectomy or watchful waiting in early prostate cancer. N Engl J Med 370(10):932–942
23. Wilt TJ, Brawer MK, Jones KM et al (2012) Radical prostatectomy versus observation for localized prostate cancer. N Engl J Med 367(3):203–213
24. Wilt TJ, Jones KM, Barry MJ et al (2017) Follow-up of prostatectomy versus observation for early prostate cancer. N Engl J Med 377(2):132–142
25. Potosky AL, Davis WW, Hoffman RM et al (2004) Five-year outcomes after prostatectomy or radiotherapy for prostate cancer: the prostate cancer outcomes study. J Natl Cancer Inst 96(18):1358–1367
26. Bul M, Zhu X, Valdagni R et al (2013) Active surveillance for low-risk prostate cancer worldwide: the PRIAS study. Eur Urol 63(4):597–603
27. Dall'Era MA, Klotz L (2017) Active surveillance for intermediate-risk prostate cancer. Prostate Cancer Prostatic Dis 20(1):1–6
28. Savdie R, Aning J, So AI, Black PC, Gleave ME, Goldenberg SL (2017) Identifying intermediate-risk candidates for active surveillance of prostate cancer. Urol Oncol 35(10):605 e601–605 e608
29. Lee H, Lee IJ, Byun SS, Lee SE, Hong SK (2017) Favorable Gleason 3 + 4 prostate cancer shows comparable outcomes with gleason 3 + 3 prostate cancer: implications for the expansion of selection criteria for active surveillance. Clin Genitourin Cancer 15(6):e1117–e1122
30. Nyame YA, Almassi N, Haywood SC et al (2017) Intermediate-term outcomes for men with very low/low and intermediate/high risk prostate cancer managed by active surveillance. J Urol 198(3):591–599
31. Network NCC (2016) NCCN guidelines version 3.2016 prostate cancer
32. Epstein JI, Walsh PC, Carmichael M, Brendler CB (1994) Pathologic and clinical findings to predict tumor extent of nonpalpable (stage T1c) prostate cancer. JAMA 271(5):368–374
33. Stamey TA, Freiha FS, McNeal JE, Redwine EA, Whittemore AS, Schmid HP (1993) Localized prostate cancer. Relationship of tumor volume to clinical significance for treatment of prostate cancer. Cancer 71(3 Suppl):933–938

34. D'Amico AV, Whittington R, Malkowicz SB et al (1998) Biochemical outcome after radical prostatectomy, external beam radiation therapy, or interstitial radiation therapy for clinically localized prostate cancer. JAMA 280(11):969–974
35. Cadeddu J (2017) Clinically localized prostate cancer: AUA/ASTRO/SUO guideline very low-/low-risk disease. Paper presented at 2017 AUA national meeting; Boston, MA
36. Tosoian JJ, Trock BJ, Landis P et al (2011) Active surveillance program for prostate cancer: an update of the Johns Hopkins experience. J Clin Oncol 29(16):2185–2190
37. Porten SP, Whitson JM, Cowan JE et al (2011) Changes in prostate cancer grade on serial biopsy in men undergoing active surveillance. J Clin Oncol 29(20):2795–2800
38. Selvadurai ED, Singhera M, Thomas K et al (2013) Medium-term outcomes of active surveillance for localised prostate cancer. Eur Urol 64(6):981–987
39. Cooperberg MR, Broering JM, Carroll PR (2010) Time trends and local variation in primary treatment of localized prostate cancer. J Clin Oncol 28(7):1117–1123
40. van den Bergh RC, Vasarainen H, van der Poel HG et al (2010) Short-term outcomes of the prospective multicentre 'prostate cancer research international: active surveillance' study. BJU Int 105(7):956–962
41. Conti SL, Dall'era M, Fradet V, Cowan JE, Simko J, Carroll PR (2009) Pathological outcomes of candidates for active surveillance of prostate cancer. J Urol 181(4):1628–1633 discussion 1633–1624
42. Suardi N, Briganti A, Gallina A et al (2010) Testing the most stringent criteria for selection of candidates for active surveillance in patients with low-risk prostate cancer. BJU Int 105(11):1548–1552
43. Alan J, Wein LRK, Partin AW, Peters CA (2016) Campbell-Walsh urology, vol 11. Elsevier, Philadelphia, PA
44. Sanda MG, Cadeddu JA, Kirkby E et al (2017) Clinically localized prostate cancer: AUA/ASTRO/SUO guideline. Part I: risk stratification, shared decision making, and care options. J Urol pii:S0022-5347(17)78003-2
45. Mohler JL, Armstrong AJ, Bahnson RR et al (2016) Prostate cancer, version 1.2016. J Natl Compr Cancer Netw 14(1):19–30
46. Dall'Era MA, Albertsen PC, Bangma C et al (2012) Active surveillance for prostate cancer: a systematic review of the literature. Eur Urol 62(6):976–983
47. Ploussard G, Nicolaiew N, Marchand C et al (2014) Prospective evaluation of an extended 21-core biopsy scheme as initial prostate cancer diagnostic strategy. Eur Urol 65(1): 154–161
48. Linder BJ, Frank I, Umbreit EC et al (2013) Standard and saturation transrectal prostate biopsy techniques are equally accurate among prostate cancer active surveillance candidates. Int J Urol 20(9):860–864
49. Taira AV, Merrick GS, Bennett A et al (2013) Transperineal template-guided mapping biopsy as a staging procedure to select patients best suited for active surveillance. Am J Clin Oncol 36(2):116–120
50. Onik G, Miessau M, Bostwick DG (2009) Three-dimensional prostate mapping biopsy has a potentially significant impact on prostate cancer management. J Clin Oncol 27(26):4321–4326
51. Lin DW, Newcomb LF, Brown MD et al (2017) Evaluating the four Kallikrein panel of the 4Kscore for prediction of high-grade prostate cancer in men in the canary prostate active surveillance study. Eur Urol 72(3):448–454
52. Eure G, Germany R, Given R et al (2017) Use of a 17-gene prognostic assay in contemporary urologic practice: results of an interim analysis in an observational cohort. Urology 107:67–75
53. Arsov C, Jankowiak F, Hiester A et al (2014) Prognostic value of a cell-cycle progression score in men with prostate cancer managed with active surveillance after MRI-guided prostate biopsy—a pilot study. Anticancer Res 34(5):2459–2466
54. Tosoian JJ, JohnBull E, Trock BJ et al (2013) Pathological outcomes in men with low risk and very low risk prostate cancer: implications on the practice of active surveillance. J Urol 190(4):1218–1222

55. Umbehr MH, Platz EA, Peskoe SB et al (2014) Serum prostate-specific antigen (PSA) concentration is positively associated with rate of disease reclassification on subsequent active surveillance prostate biopsy in men with low PSA density. BJU Int 113(4):561–567
56. Ross AE, Loeb S, Landis P et al (2010) Prostate-specific antigen kinetics during follow-up are an unreliable trigger for intervention in a prostate cancer surveillance program. J Clin Oncol 28(17):2810–2816
57. Dall'Era MA, Konety BR, Cowan JE et al (2008) Active surveillance for the management of prostate cancer in a contemporary cohort. Cancer 112(12):2664–2670
58. Xia J, Trock BJ, Cooperberg MR et al (2012) Prostate cancer mortality following active surveillance versus immediate radical prostatectomy. Clin Cancer Res 18(19):5471–5478
59. Klotz L (2012) Active surveillance: the Canadian experience with an "inclusive approach". J Natl Cancer Inst Monogr 2012(45):234–241
60. Bokhorst LP, Valdagni R, Rannikko A et al (2016) A decade of active surveillance in the PRIAS study: an update and evaluation of the criteria used to recommend a switch to active treatment. Eur Urol 70(6):954–960
61. Oberlin DT, Casalino DD, Miller FH, Meeks JJ (2017) Dramatic increase in the utilization of multiparametric magnetic resonance imaging for detection and management of prostate cancer. Abdom Radiol (NY) 42(4):1255–1258
62. Almeida GL, Petralia G, Ferro M et al (2016) Role of multi-parametric magnetic resonance image and PIRADS score in patients with prostate cancer eligible for active surveillance according PRIAS criteria. Urol Int 96(4):459–469
63. Glaser ZA, Gordetsky JB, Porter KK, Varambally S, Rais-Bahrami S (2017) Prostate cancer imaging and biomarkers guiding safe selection of active surveillance. Front Oncol 7:256
64. Grey AD, Chana MS, Popert R, Wolfe K, Liyanage SH, Acher PL (2015) Diagnostic accuracy of magnetic resonance imaging (MRI) prostate imaging reporting and data system (PI-RADS) scoring in a transperineal prostate biopsy setting. BJU Int 115(5):728–735
65. Yim JH, Kim CK, Kim JH (2017) Clinically insignificant prostate cancer suitable for active surveillance according to prostate cancer research international: active surveillance criteria: utility of PI-RADS v2. J Magn Reson Imaging 47(4):1072–1079
66. Porpiglia F, Cantiello F, De Luca S et al (2016) Multiparametric magnetic resonance imaging and active surveillance: how to better select insignificant prostate cancer? Int J Urol 23(9):752–757
67. Hoeks CM, Somford DM, van Oort IM et al (2014) Value of 3-T multiparametric magnetic resonance imaging and magnetic resonance-guided biopsy for early risk restratification in active surveillance of low-risk prostate cancer: a prospective multicenter cohort study. Investig Radiol 49(3):165–172
68. Stamatakis L, Siddiqui MM, Nix JW et al (2013) Accuracy of multiparametric magnetic resonance imaging in confirming eligibility for active surveillance for men with prostate cancer. Cancer 119(18):3359–3366
69. Lai WS, Gordetsky JB, Thomas JV, Nix JW, Rais-Bahrami S (2017) Factors predicting prostate cancer upgrading on magnetic resonance imaging-targeted biopsy in an active surveillance population. Cancer 123(11):1941–1948
70. Lim CS, McInnes MDF, Flood TA et al (2017) Prostate imaging reporting and data system, version 2, assessment categories and pathologic outcomes in patients with Gleason score 3 + 4 = 7 prostate cancer diagnosed at biopsy. AJR Am J Roentgenol 208(5):1037–1044
71. Nougaret S, Robertson N, Golia Pernicka J et al (2017) The performance of PI-RADSv2 and quantitative apparent diffusion coefficient for predicting confirmatory prostate biopsy findings in patients considered for active surveillance of prostate cancer. Abdom Radiol (NY) 42(7):1968–1974
72. Tan WP, Mazzone A, Shors S et al (2017) Central zone lesions on magnetic resonance imaging: should we be concerned? Urol Oncol 35(1):31 e37–31 e12
73. Panebianco V, Barchetti F, Sciarra A et al (2015) Multiparametric magnetic resonance imaging vs. standard care in men being evaluated for prostate cancer: a randomized study. Urol Oncol 33(1):17 e11–17 e17

74. Da Rosa MR, Milot L, Sugar L et al (2015) A prospective comparison of MRI-US fused targeted biopsy versus systematic ultrasound-guided biopsy for detecting clinically significant prostate cancer in patients on active surveillance. J Magn Reson Imaging 41(1):220–225
75. Nassiri N, Margolis DJ, Natarajan S et al (2017) Targeted biopsy to detect Gleason score upgrading during active surveillance for men with low versus intermediate risk prostate cancer. J Urol 197(3 Pt 1):632–639
76. Abdi H, Pourmalek F, Zargar H et al (2015) Multiparametric magnetic resonance imaging enhances detection of significant tumor in patients on active surveillance for prostate cancer. Urology 85(2):423–428
77. Weaver JK, Kim EH, Vetter JM, Fowler KJ, Siegel CL, Andriole GL (2016) Presence of magnetic resonance imaging suspicious lesion predicts Gleason 7 or greater prostate cancer in biopsy-naive patients. Urology 88:119–124
78. Marliere F, Puech P, Benkirane A et al (2014) The role of MRI-targeted and confirmatory biopsies for cancer upstaging at selection in patients considered for active surveillance for clinically low-risk prostate cancer. World J Urol 32(4):951–958
79. Frye TP, George AK, Kilchevsky A et al (2017) Magnetic resonance imaging-transrectal ultrasound guided fusion biopsy to detect progression in patients with existing lesions on active surveillance for low and intermediate risk prostate cancer. J Urol 197(3 Pt 1):640–646
80. Siddiqui MM, Truong H, Rais-Bahrami S et al (2015) Clinical implications of a multiparametric magnetic resonance imaging based nomogram applied to prostate cancer active surveillance. J Urol 193(6):1943–1949
81. Felker ER, Wu J, Natarajan S et al (2016) Serial magnetic resonance imaging in active surveillance of prostate cancer: incremental value. J Urol 195(5):1421–1427
82. Rais-Bahrami S, Turkbey B, Rastinehad AR et al (2014) Natural history of small index lesions suspicious for prostate cancer on multiparametric MRI: recommendations for interval imaging follow-up. Diagn Interv Rad (Ankara) 20(4):293–298
83. Moore CM, Giganti F, Albertsen P et al (2017) Reporting magnetic resonance imaging in men on active surveillance for prostate cancer: the PRECISE recommendations—a report of a European School of Oncology Task Force. Eur Urol 71(4):648–655
84. Walton Diaz A, Shakir NA, George AK et al (2015) Use of serial multiparametric magnetic resonance imaging in the management of patients with prostate cancer on active surveillance. Urol Oncol 33(5):202.e201–202.e207
85. Weinreb JC, Barentsz JO, Choyke PL et al (2016) PI-RADS prostate imaging – reporting and data system: 2015, version 2. Eur Urol 69(1):16–40
86. Scheltema MJ, Tay KJ, Postema AW et al (2017) Utilization of multiparametric prostate magnetic resonance imaging in clinical practice and focal therapy: report from a Delphi consensus project. World J Urol 35(5):695–701
87. Arumainayagam N, Ahmed HU, Moore CM et al (2013) Multiparametric MR imaging for detection of clinically significant prostate cancer: a validation cohort study with transperineal template prostate mapping as the reference standard. Radiology 268(3):761–769
88. Russo F, Regge D, Armando E et al (2016) Detection of prostate cancer index lesions with multiparametric magnetic resonance imaging (mp-MRI) using whole-mount histological sections as the reference standard. BJU Int 118(1):84–94
89. Rais-Bahrami S, Siddiqui MM, Turkbey B et al (2013) Utility of multiparametric magnetic resonance imaging suspicion levels for detecting prostate cancer. J Urol 190(5):1721–1727
90. Rais-Bahrami S, Siddiqui MM, Vourganti S et al (2015) Diagnostic value of biparametric magnetic resonance imaging (MRI) as an adjunct to prostate-specific antigen (PSA)-based detection of prostate cancer in men without prior biopsies. BJU Int 115(3):381–388
91. Moore CM, Petrides N, Emberton M (2014) Can MRI replace serial biopsies in men on active surveillance for prostate cancer? Curr Opin Urol 24(3):280–287
92. Giganti F, Moore CM, Robertson NL et al (2017) MRI findings in men on active surveillance for prostate cancer: does dutasteride make MRI visible lesions less conspicuous? Results from a placebo-controlled, randomised clinical trial. Eur Radiol 27(11):4767–4774

93. Siddiqui MM, Rais-Bahrami S, Turkbey B et al (2015) Comparison of MR/ultrasound fusion-guided biopsy with ultrasound-guided biopsy for the diagnosis of prostate cancer. JAMA 313(4):390–397
94. Nahar B, Katims A, Barboza MP et al (2017) Reclassification rates of patients eligible for active surveillance after the addition of magnetic resonance imaging-ultrasound fusion biopsy: an analysis of 7 widely used eligibility criteria. Urology 110:134–139
95. Sonn GA, Filson CP, Chang E et al (2014) Initial experience with electronic tracking of specific tumor sites in men undergoing active surveillance of prostate cancer. Urol Oncol 32(7):952–957
96. Palapattu GS, Salami SS, Cani AK et al (2017) Molecular profiling to determine clonality of serial magnetic resonance imaging/ultrasound fusion biopsies from men on active surveillance for low-risk prostate cancer. Clin Cancer Res 23(4):985–991
97. Okoro C, George AK, Siddiqui MM et al (2015) Magnetic resonance imaging/transrectal ultrasonography fusion prostate biopsy significantly outperforms systematic 12-core biopsy for prediction of total magnetic resonance imaging tumor volume in active surveillance patients. J Endourol 29(10):1115–1121
98. Ristau BT, Chen DYT, Ellis J et al (2017) Defining novel and practical metrics to assess the deliverables of multiparametric magnetic resonance imaging/ultrasound fusion prostate biopsy. J Urol 199(4):969–975
99. Chang E, Jones TA, Natarajan S et al (2017) Value of tracking biopsy in men undergoing active surveillance of prostate cancer. J Urol 199(1):98–105
100. Eggener SE, Mueller A, Berglund RK et al (2013) A multi-institutional evaluation of active surveillance for low risk prostate cancer. J Urol 189(1 Suppl):S19–S25 discussion S25

Chapter 4
Evaluation of Prostate Needle Biopsies

Giovanna A. Giannico and Omar Hameed

Abstract The introduction of Prostate Specific Antigen (PSA) screening has caused a stage shift in diagnosis of prostate cancer and an increasing number of patients receiving early diagnosis. This has led to early detection of limited foci of cancer on prostate biopsy, and clinically insignificant prostate cancer at radical prostatectomy. Therefore, current methods for sampling, diagnosing and managing prostate cancer have significantly evolved in recent years. In light of recent management changes and conservative surveillance protocols prompting new handling, grading and staging guidelines, the evaluation of prostate biopsy in contemporary practice has become pivotal. It is therefore critical to recognize minor foci of cancer or atypical glands, and distinguish these from benign mimics that could lead to a false positive diagnosis.

In this chapter, current biopsy modalities and imaging techniques, tissue handling and recent updates in the interpretation of prostate biopsy will be discussed.

Keywords Prostate cancer · Prostate needle biopsy · Gleason score · Pathology prostate biopsy · Prostate biopsy diagnosis

4.1 Indications for Prostate Biopsy and Sampling Techniques

Prostate cancer is the most common cancer in the US and the second highest cause of death expected to occur in men in 2018 [1], and its incidence is increasing due to the use of screening methods. Diagnosis of prostate cancer is established by prostate needle biopsies. Serum prostate-specific antigen (PSA) and digital rectal

G. A. Giannico (✉)
Pathology Medical Director, HCA Midwest Division, Kansas City, MO, USA

O. Hameed
Adjunct Professor of Pathology, Microbiology and Immunology, Vanderbilt University Medical Center, Nashville, TN, USA

© Springer Nature Switzerland AG 2018
H. Schatten (ed.), *Molecular & Diagnostic Imaging in Prostate Cancer*,
Advances in Experimental Medicine and Biology 1096,
https://doi.org/10.1007/978-3-319-99286-0_4

examination (DRE) are current methods utilized to identify patients at risk of prostate cancer who are candidates for biopsy.

Biopsy modalities have evolved with time. The National Comprehensive Cancer Network (NCCN) guidelines recommend transrectal ultrasound (TRUS)-guided biopsy as sampling modality. Initially introduced as sextant biopsy in 1989 with sampling of apex, mid and base of bilateral prostate lobes in the mid parasagittal plane [2], this protocol has subsequently evolved due to false-negative rates exceeding 30%. Current standard of care in sampling modality is a 12-core extended sextant biopsy [3, 4]. This protocol, which targets the peripheral zone of the prostate, an area that harbors approximately 70–80% of all prostate cancers, significantly enhances diagnosis of cancer by lateral sampling in addition to the standard parasagittal sampling of the original sextant protocol. Sampling of transition and anterior zone in cases with persistently elevated PSA may also be considered in addition to the standard protocol. Of note, however, studies on the utility of additional anterior sampling have demonstrates only marginal improvement in cancer detection rates [5], and performance in detecting prostate cancer is not significantly increased by sampling >12 cores [3, 6–8].

In an attempt to decrease the rate of false negative biopsies, more extensive biopsy schemes have been proposed. Saturation [9, 10], transrectal or transperineal biopsies aim to improve cancer detection rates in patients at increased risk of prostate cancer with previously negative biopsies [11] and accurately predict tumor volume and grade in patients with known prostate cancer compared with traditional biopsy schemes. This is carried out by sampling >20 cores with a threshold of 22–24 cores arbitrarily set [12]. Transperineal template-guided mapping biopsy uses a standard brachytherapy grid with holes 5-mm apart as a template [13–15].

In recent years, multiparametric magnetic resonance imaging (MRI) targeted prostate biopsy has changed the standard practice of prostate cancer sampling. However, to date the rationale for its use still awaits validation with clinical trials. In men with an elevated serum PSA and a prior negative TRUS biopsy, MRI targeted biopsy has been shown to decrease the number of repeat biopsies, improve detection of significant cancers, and decrease the number of biopsy cores [16–22]. Current NCCN guidelines support the use of MRI targeted biopsy in patients with at least one prior negative biopsy.

4.2 Pathologic Evaluation of Prostate Biopsies

The prostate consists of acini composed by glandular structures in a lobular pattern embedded within fibromuscular stroma. The acini connect to secretory ducts, lined by a low cuboidal epithelium, which becomes transitional when these ducts open into the urethra. The glands are lined by secretory cells and are surrounded by basal cells.

4.2.1   Major Prostatic Morphologic Lesions

Prostatic Adenocarcinoma. The differential diagnosis of prostatic adenocarcinoma from benign conditions, particularly in the setting of limited carcinoma may be challenging. Major criteria for diagnosis of prostate cancer include: (1) Small glands with infiltrative pattern and cribriform glands that are too large and/or irregular to represent high grade prostatic intraepithelial neoplasia (HGPIN), a precursor lesion of prostatic adenocarcinoma; (2) Nuclear enlargement and nucleolar prominence; (3) Single cell layer (loss of basal cells). Minor criteria include: (1) Blue-tinged mucinous secretions; (2) Pink amorphous secretions; (3) Mitotic figures; (4) Crystalloids; (5) Adjacent high grade PIN; (6) Amphophilic cytoplasm; (7) Nuclear hyperchromasia [23] (Fig. 4.1). Perineural invasion, mucinous fibroplasia and

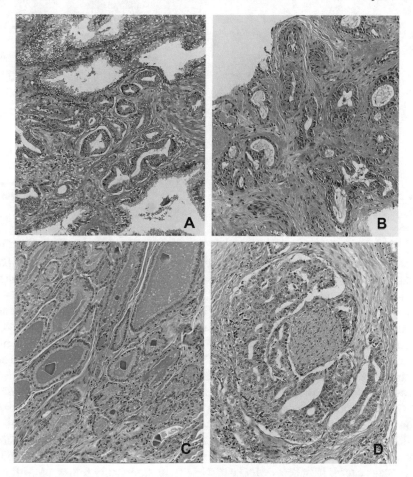

Fig. 4.1 Criteria for diagnosis of prostatic adenocarcinoma: (**a**) Infiltrative pattern of prostatic adenocarcinoma among benign glands. Note the presence of amphophilic cytoplasm compared to the adjacent benign glands, nuclear enlargement and nucleolar prominence; (**b**) Blue mucin within lumina of neoplastic glands; (**c**) Pink amorphous secretions and crystalloids; (**d**) Perineural invasion (Original magnification, 10×)

Table 4.1 Histologic mimics of prostate cancer

Anatomic structures
Benign crowded glands
Seminal vesicle/ejaculatory duct
Cowper's gland
Paraganglion
Verumontanum mucosal glands
Mesonephric gland remnants
Clear cell cribriform hyperplasia
Atrophy
Partial atrophy
Post-atrophic hyperplasia
Reactive atypia
Inflammation
Radiation
Benign prostatic glands with radiation effect
Metaplasia
Mucinous metaplasia
Urothelial metaplasia
Benign glandular proliferations
Nephrogenic (adenoma)
Basal cell hyperplasia
Adenosis
Sclerosing adenosis

Adapted from: Srigley JR. Benign mimickers of prostate cancer

glomerulations are diagnostic features that are specific for prostate cancer [24]. A comprehensive list of benign entities that enter in the diagnosis of prostatic adenocarcinoma are listed in Table 4.1 [25, 26].

High Grade Prostatic Intraepthelial Neoplasia (HGPIN). HGPIN is diagnosed when cytologic changes similar those seen in adenocarcinoma and visible at 20× magnification [27] are identified in architecturally benign glands. HGPIN is a morphologic precursor of prostate cancer, and is found concurrently with prostatic adenocarcinoma in up to 85% of cases [28]. Isolated HGPIN can be identified in 5–8% of cases in prostate biopsies without concurrent adenocarcinoma [29]. In this setting, the risk of prostatic adenocarcinoma in repeat biopsy is similar to that of the general population, i.e. 24.1% [29]. However, several studies have shown that multifocal HGPIN, i.e. detected in two or more biopsy cores nearly doubles the risk of cancer in a subsequent biopsy [30–33]. In view of radical prostatectomy data showing low grade/stage cancer in patients with HGPIN [34], multimodality follow-up with multiparametric MRI, molecular markers and repeat biopsy at 1 year in selected cases, similarly to patients in active surveillance, has been recommended [35].

Intraductal Carcinoma (IDC). IDC of the prostate, first described in 1973 [36], is the defined by the presence of malignant cells involving pre-existing prostatic ducts and acini and surrounded by basal cells. Criteria for morphologic diagnosis of IDC include an intraductal proliferation of malignant cells with solid or densely cribriform pattern and involving >70% of the ductal/acinar space; in absence of these features, a loose cribriform and/or micropapillary proliferation requires marked nuclear atypia (6× normal) or necrosis [37]. IDC is associated with high grade prostate cancer [38], and is an independent prognostic factor for recurrence and mortality [39, 40]. When associated with pattern 4 cribriform adenocarcinoma, which often coexists with IDC, it is an independent predictor of biochemical recurrence after prostatectomy, while the percent of pattern 4 adenocarcinoma is not [41]. Furthermore, in a recent study, the incidence of IDC was strongly associated with increasing National Comprehensive Cancer Network (NCCN) risk classes: 2.1% for low risk, 23.1% for intermediate risk, 36.7% for high risk, and 56.0% for metastatic disease [42]. Intraductal spread of cancer within ducts and acini in a retrograde manner should be distinguished from "precursor IDC", which represents a de novo intraductal lesion, unassociated with invasive cancer [43].

4.3   Immunohistochemistry

Although the diagnosis of prostatic adenocarcinoma is based on evaluation of a combination of cytologic and architectural features, as discussed above, morphology alone may be insufficient, especially in the setting of limited cancer, and immunohistochemistry is commonly used to distinguish adenocarcinoma from benign mimickers when evaluating prostate biopsies. The International Society of Urologic Pathology (ISUP), the international professional organization for uropathology, recommends the use of high molecular weight cytokeratin (HMWCK) such as CK5/6 or 34[beta]E12 and p63 as specific markers for basal cells, and alpha-methylacyl-CoA racemase (AMACR) singly or in various cocktail combination [44] (Fig. 4.2). Although typically adenocarcinoma lacks expression of basal cell markers, benign lesions such as adenosis, atrophy, or benign glands may also demonstrate similar basal cell loss [45–48]. Conversely, HMWCK staining in a non-basal distribution and aberrant diffuse expression of p63 may occasionally be observed in prostate cancer [49–51]. Additionally, AMACR may be positive in 5–21% of benign prostatic glands, [45, 47, 52, 53] and up to 18% of cases of adenosis, [54] which limits its specificity for the diagnosis of adenocarcinoma. ETS-related gene (ERG), a member of the erythroblast transformation-specific (ETS) family of transcriptions factors, located on chromosome 21q22.2 is suggested as an optional marker by the ISUP, due to low sensitivity and intratumoral heterogeneity [44] in view of the relatively frequent expression in foci of HGPIN adjacent to adenocarcinoma.

Another clinical setting in which immunohistochemistry may be very useful is in the differential diagnosis between primary adenocarcinoma of the prostate and secondary malignancy involving the prostate by adjacent spread or metastasis.

Fig. 4.2 Immunohistochemistry with an antibody cocktail of two basal cell markers (p63, nuclear and HMWCK, cytoplasmic, brown) and a cytoplasmic marker preferentially expressed in prostatic adenocarcinoma (AMACR, red). Basal cells are present in benign glands and are lost in foci of adenocarcinoma (Original magnification, 10×)

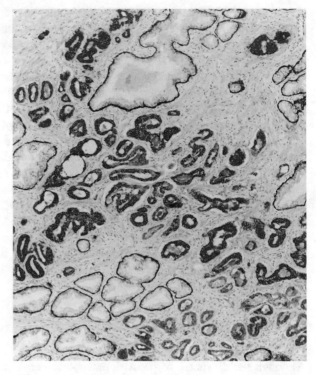

PSA and PSAP have been widely used in the past. However, poorly high grade/poorly differentiated prostatic adenocarcinomas may not express PSA, and androgen deprivation treatment may decrease PSA secretion as well. P501S, PSMA and NKX3.1 are recommended markers that may be performed when staining for PSA is equivocal [44].

4.4　Grading of Prostatic Adenocarcinoma

The current histologic grading system was developed by Donald Gleason in 1966 [55]. This system assigned a grade on a scale from 1 to 5 based on the two most dominant histologic architectural patterns observed at low power (4× or 10×). The introduction of PSA screening since the late 1980s has changed the current practices of diagnosing and treating prostate cancer with a striking decline in the rate of metastatic cancer by 50% [56] and with a 20% decline in prostate cancer deaths [57]. This has prompted the need for a revision of the grading system as originally developed by Gleason. Specifically, current practices in the era of "insignificant "tumors require the use of extended biopsy protocols compared to a few biopsy cores in the Gleason's era of palpable tumors. Gleason's work was based on morphology alone prior to the introduction of immunohistochemistry for detection of basal cells, with some of the originally described grading patterns 1 and 2 likely representing adenosis or partial atrophy [58]. Additionally, recent studies have high-lighted the adverse outcome of cribriform pattern, originally classified as pattern 3, requiring a grade shift of cancers with such morphology toward pattern 4 or intra-ductal carcinoma [40, 59]. Furthermore, the role of tertiary grades was not addressed by Gleason in his seminal work. Revisions of the original Gleason grading system occurred in 2005 and, more recently in 2014 based on the work of the ISUP (Figs. 4.3 and 4.4). In the most recent 2014 consensus conference, which included pathologists, urologists, and urologic medical and radiation oncologists, a multidis-ciplinary update on the Gleason Grading system was proposed. This was subse-quently endorsed by the World Health Organization (WHO) Classification of Tumors of the Urinary and Male Reproductive System and the American Joint Committee on Cancer (AJCC) TNM staging manual. Important changes in the Gleason grading system and grading of variant morphologies are illustrated in Table 4.2 and Figs. 4.3 and 4.4. An important change discussed in the 2014 ISUP Consensus Conference was the introduction of a new grading system. The original and 2005 modified Gleason grading systems were assigned based on morphologic criteria. The new Grade Group grading system, developed utilizing grade stratifica-tion based on prognostic significance rather than morphology, introduces categories

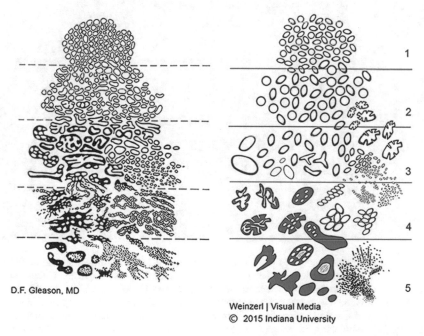

Fig. 4.3 Diagrams of morphologic patterns of prostatic adenocarcinoma. Original Gleason's description (left panel), and 2014 modified International Society of Urologic Pathology grading (right panel). Reprinted from Epstein et al. [68] with permission from Wolters Kluwer Health, Inc. Copyright 2017 Copyright Clearance Center Inc. All permission requests for this image should be made to the copyright holder

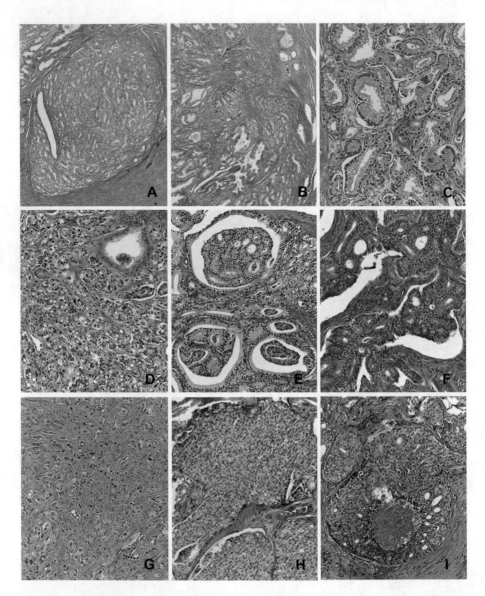

Fig. 4.4 Morphologic features of prostate cancer grading. (**a**) Circumscribed nodule of Gleason pattern 1 with back-to back neoplastic glands with uniform size and shape; (**b**) Circumscribed nodule of Gleason pattern 2. Note mild variation in shape and size of the glands; (**c**) Gleason pattern 3 adenocarcinoma with regularly spaced although dense small well-formed glands; (**d**) Poorly formed glands of Gleason pattern 4; (**e**) Glomeruloid pattern 4. Note intraluminal projections of tumor nodules mimicking a renal glomerulus; (**f**) Cribriform pattern 4; (**g**) Single cell pattern of Gleason 5; (**h**) Solid sheets of Gleason pattern 5; (**i**) Comedonecrosis in Gleason pattern 5 (Original magnification, 10×)

Table 4.2 Modified ISUP Gleason grading systems

2005 ISUP modified Gleason grading system	2014 ISUP modified Gleason grading system
Gleason pattern 1	
Should not be diagnosed: most cases likely represent adenosis	
Gleason pattern 2	
Should rarely, if ever diagnosed on biopsy	
Gleason pattern 3	
Excludes individual cells	
Excludes cribriform glands with the exception of well-circumscribed cribriform glands of the same size of normal glands with regular contour and round evenly spaced lumina	Excludes all cribriform glands
Gleason pattern 4	
Ill-defined glands with poorly formed glandular lumina where a tangential section of Gleason pattern 3 glands cannot account for the histology	
Cribriform glands	All cribriform glands, regardless of morphology should be assigned a Gleason pattern 4
Fused glands	
No consensus on grading of glomeruloid glands	Glomeruloid glands regardless of morphology should be assigned a Gleason pattern 4
Gleason pattern 5	
Solid sheets, cord or single cells	
Comedonecrosis	
Grading variants	
Tumors with vacuoles are distinct from true signet ring cells and should be graded based on underlying architectural patterns	
Foamy gland carcinoma should be graded based on underlying architectural patterns	
Ductal adenocarcinoma should be graded as Gleason score 4 + 4 = 8. In cases with mixed ductal and acinar patterns, the ductal patterns should be assigned Gleason pattern 4	
Colloid (mucinous) carcinoma with cribriform glands should be graded as 4 + 4 = 8. Grading of cases with discrete individual glands was controversial and may be graded as 4 + 4 = 8 or 3 + 3 = 6	Grading of mucinous carcinoma should be based on its underlying growth pattern rather without defaulting to pattern 4
Small cell carcinoma should not be assigned a Gleason grade	

(continued)

Table 4.2 (continued)

2005 ISUP modified Gleason grading system	2014 ISUP modified Gleason grading system
Focal mucinous extravasation should be ignored and the tumor should be graded based on the underlying glandular architecture	
Mucinous fibroplasia should subtract away tumor should be graded based on the underlying glandular architecture	
The grading of glomeruloid structures was controversial and may be graded as 4 + 4 = 8 or 3 + 3 = 6	
Pseudohyperplastic adenocarcinoma should be graded as Gleason score 3 + 3 = 6	
Grading of needle core biopsies	
In the setting of high-grade cancer, lower-grade patterns should be ignored if <5% of the tumor	
Any amount of high grade tumor should be included in the Gleason grade. Consequently, tertiary grades are dropped and primary pattern and the highest grade should be recorded	
Assign individual Gleason scores to separate cores submitted in separate containers or in the same container with individual site designation (ie, by different color inks)	
No consensus on grading different cores with different grades from same specimen container without site designation	
Give an overall score for multiple fragmented cores in the same container	
Reporting percent pattern 4/5 is optional	Reporting percent pattern 4 is recommended
Grading of radical prostatectomy	
In the setting of high-grade cancer, lower-grade patterns should be ignored if <5% of the tumor	
Gleason score based on the primary and secondary patterns	
Assign a separate Gleason score to each dominant tumor nodule(s) with a comment on the tertiary pattern	
Reporting percent pattern 4/5 is optional	Reporting percent pattern 4 is recommended

(continued)

Table 4.2 (continued)

2005 ISUP modified Gleason grading system	2014 ISUP modified Gleason grading system
Tertiary pattern	
Not applicable to biopsies	
Report in radical prostatectomies	The preferred term for tertiary pattern is "minor high grade pattern". It should only be used in the logical scenario when there are 3 grade patterns, such as with 3 + 4 = 7 or 4 + 3 = 7 with <5% Gleason pattern 5 at radical prostatectomy. No cut off addressed
Intraductal carcinoma	
Not discussed	Intraductal carcinoma should not be assigned a Gleason grade

ISUP International Society of Urologic Pathology

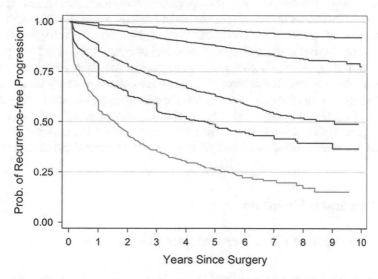

Fig. 4.5 New prognostic prostatic adenocarcinoma grade groups and biochemical recurrence-free survival. Reprinted from Epstein et al. [68] with permission from Wolters Kluwer Health, Inc. Copyright 2017 Copyright Clearance Center Inc. All permission requests for this image should be made to the copyright holder

from 1 to 5 based on Gleason grading and outcome data of biochemical recurrence free survival (Fig. 4.5). Therefore, Grade Groups 1–5 refer to Gleason scores ≤6, 3 + 4 = 7, 4 + 3 = 7, 4 + 4 = 8 and grades 9–10 with a corresponding BCR-free survival after RP of 97%, 88%, 70%, 64%, and 34%, respectively. This grading system was developed based on data from 7869 patients [60], and recently validated on a larger multicenter study [61]. The updated grading system has been incorporated into the recently updated College of American Pathologists (CAP) Cancer Protocols and the 8th edition of the AJCC staging manual.

4.5 The Concept of "Limited Carcinoma"

Despite the persistent controversy about PSA screening effect on mortality, as emerged with the contrasting results of the Prostate, Lung, Colorectal, and Ovarian Cancer Screening Trial [62] and the European Randomized Study of Screening for Prostate Cancer [57] trials, the former showing no difference in mortality with PSA screening, the latter suggesting reduced prostate cancer mortality by 20%, the most significant effect of PSA screening has been that of a stage migration. However, this has led to overdiagnosis and overtreatment of potentially indolent ("insignificant") cancer at low risk for progression [63]. In recent years, active surveillance (AS) of prostate cancer with PSA monitoring and periodic repeat biopsies has emerged as a potential management strategy for indolent cancer with the intent of avoiding unnecessary radical surgical treatment. Risk stratification for inclusion in AS protocols is based both on clinical and pathologic characteristics such as PSA, clinical stage, pathologic grading, number of biopsy cores with cancer, and extent of cancer in any core. Therefore, histologic evaluation of prostate biopsy has a central role in management of patients with prostate cancer. However, histologic evaluation becomes challenging in the setting of minute foci of cancer ("minimal" or "limited" adenocarcinoma), where differentiation from benign mimickers becomes crucial. "Minimal" or "limited" adenocarcinoma has been defined as the presence of few malignant glands on biopsy, measuring less than 1 mm in length and occupying less than 5% of needle core tissue [64, 65]. A minimal number of 2–20 glands has also been suggested [23, 26, 64–66] for this definition. Limited carcinoma on biopsy does not necessarily equate with insignificant cancer on radical prostatectomy.

4.6 Specimen Handling

4.6.1 Number of Cores per Specimen Container

The number of biopsy cores submitted in each individual specimen container greatly affects the quality of tissue processing and subsequent pathologic interpretation. Tissue may be received as single cores in 12 or more site-specific individually labeled containers, or in six containers, each containing two cores representing sextant sites, or in two containers each containing six cores representing the right and left sides. Several issues are associated with single or multiple core placement in individual jars. Current active surveillance protocols require core-specific quantification of tumor grading and extent of individual core involvement, which may be best and more accurately evaluated when individual cores are separately submitted in site-specific labeled containers. When multiple cores are submitted in the same container without site-specific designation, individual grading of each core or global grading of the entire specimen could be applied. However, issues with both grading approaches were discussed at the 2005 consensus meeting, and in a recent update to the 2014 ISUP meeting [58, 67, 68]. Specifically, applying a separate grading in multiple biopsy cores has significant repercussions in the event, e.g. of one core showing $3 + 3 = 6$ and another core showing $4 + 4 = 8$. Assigning a specific grade to individual cores would diagnose $4 + 4 = 8$, while assigning an overall grade would diagnose $3 + 4 = 7$ or $4 + 3 = 7$ with significant undergrading the final specimen as a result. Polling at the 2014 ISUP conference showed that 45.2% of participants reported each positive core, 17.7% reported each positive specimen jar, and 3.2% reported the average grade for the entire case. In view of previous studies showing different prognostic significance with grading individual cores versus assigning an overall grade, the recommendation from the 2014 update to the meeting was to "assign individual Gleason scores to separate cores as long as the cores are submitted in separate containers or the cores are in the same container yet specified by the urologist as to their location, i.e. by different color inks. In cases where there are different undesignated cores with different grades in the same specimen container, it is optional whether to assign individual grades to different cores or a global grade for the specimen container" [58, 67, 68].

Another issue related to the modality of tissue submission is that of fragmentation. Fragmentation may be the result of numerous factors. Operator expertise (inclusion of periprostatic adipose tissue when biopsy is aimed at outer peripheral zone), tissue characteristics (cystically dilated glands), transportation (entrapment of tissue in biopsy sponges), gross handling or embedding in the pathology laboratory are among the most common issues that may affect fragmentation. Fragmentation has also been directly associated with the number of cores per container [69, 70], and equivocal diagnoses have been less frequently rendered with specimens submitted in 6–12 containers compared with those submitted in 1–2 containers [71]. Of note, an average length of 1.15 cm of tissue, corresponding to an average-length prostate biopsy may be lost during processing when including three biopsy cores in the same cassette [72]. Multiple biopsies in the same container may also result in loss of 40% of the tissue surface area with only a 5-degree shift in the angle of the needle biopsy within the tissue block [73].

Complementary to the issue of fragmentation is that of site-specific individual core submission. Previous studies have shown that the rate of detecting carcinoma in the same sextant site is between 48% and 57% and up to 85% if adjacent sextant sites are included when targeting the same area(s) with prior diagnosis of atypia suspicious for carcinoma [12]. Furthermore, in recent years and with the availability of focal therapy (a recently emerging ablation treatment of the dominant or index lesion for localized prostate cancer) location of cancer site becomes more critical [74]. As a result, site-specific submission of prostate tissue may allow for risk stratification and management during follow-up of men on active surveillance or undergoing conservative treatment strategies [75] (see below for details). Although recognizing that single-core site-specific labeled submission is ideal, the CAP reports that "2 core submission is acceptable" (Version 4.0.0.0), an approach that is also endorsed by the American Urological Association.

4.6.2 Length of Biopsy Core

The length of biopsy core is an important factor for cancer detection and assessment of percent of core involvement. Several studies have addressed the impact of longer cores on cancer detection [76–79], and an average length of prostatic needle biopsies measured on glass slide >10 mm has been proposed among quality indicators [80]. Core lengths of at least 11.9 and 13 mm have been shown to represent the best cutoff for quality assurance [81].

4.7 Prostate Biopsy Pathologic Reporting

The CAP has recently updated the specimen reporting guidelines (Version 4.0.0.0) to incorporate recent changes from the 2014 ISUP consensus conference and 2016 WHO classification. Reporting of individual cores when more than one core is submitted in the same jar is recommended if the cores are individually labeled, as mentioned above. When multiple unlabeled cores are submitted in the same jar, individual or aggregate reporting of cores is optional.

Gleason score should be provided for all carcinomas with the exception of cases with treatment effect. The Gleason score represents the sum of the most predominant pattern and the second most predominant pattern. Tertiary patterns are not provided when grading needle core biopsies. Rather, a tertiary pattern with a higher grade is reported as secondary pattern independently of the tumor amount. Conversely, if a minor secondary pattern of lower grade tumor should be dismissed. When only one pattern is present, the final score should double the grade present. The new grade grouping should accompany the Gleason grade system.

Reporting percent of pattern 4 for cases with Gleason grades 3 + 4 = 7 and 4 + 3 = 7 is also required. Recording the percent of pattern 4 in grade groups higher that 3 or the percent pattern 5 is optional. The CAP also recommends reporting the number of positive cores. This does not apply with core fragmentation, where an accurate count cannot be provided. Furthermore, the linear length and percent of core involvement by tumor, including intervening stroma for discontinuous foci of tumor should be recorded. Finally, the biopsy should indicate presence of extraprostatic extension, perineural invasion and HGPIN.

4.8 Conclusions

The introduction of PSA screening has significantly modified the approach to diagnosis and management of prostate cancer but overdiagnosis and overtreatment of "insignificant" prostate cancer remain a problem. This is particularly important in view of the uncertain benefit of PSA screening on mortality. Future studies focusing on understanding the genomic signature of aggressiveness of prostate cancer are warranted.

References

1. Siegel RL, Miller KD, Jemal A (2018) Cancer statistics, 2018. CA Cancer J Clin 68:7–30
2. Hodge KK, McNeal JE, Terris MK et al (1989) Random systematic versus directed ultrasound guided transrectal core biopsies of the prostate. J Urol 142:71–74 discussion 74–75
3. Eichler K, Hempel S, Wilby J et al (2006) Diagnostic value of systematic biopsy methods in the investigation of prostate cancer: a systematic review. J Urol 175:1605–1612
4. Serefoglu EC, Altinova S, Ugras NS et al (2013) How reliable is 12-core prostate biopsy procedure in the detection of prostate cancer? Can Urol Assoc J 7:E293–E298
5. Guichard G, Larre S, Gallina A et al (2007) Extended 21-sample needle biopsy protocol for diagnosis of prostate cancer in 1000 consecutive patients. Eur Urol 52:430–435
6. Bjurlin MA, Carter HB, Schellhammer P et al (2013) Optimization of initial prostate biopsy in clinical practice: sampling, labeling and specimen processing. J Urol 189:2039–2046
7. de la Taille A, Antiphon P, Salomon L et al (2003) Prospective evaluation of a 21-sample needle biopsy procedure designed to improve the prostate cancer detection rate. Urology 61:1181–1186
8. Meng MV, Elkin EP, DuChane J et al (2006) Impact of increased number of biopsies on the nature of prostate cancer identified. J Urol 176:63–68 discussion 69
9. Stewart CS, Leibovich BC, Weaver AL et al (2001) Prostate cancer diagnosis using a saturation needle biopsy technique after previous negative sextant biopsies. J Urol 166:86–91 discussion 91–82
10. Borboroglu PG, Comer SW, Riffenburgh RH et al (2000) Extensive repeat transrectal ultrasound guided prostate biopsy in patients with previous benign sextant biopsies. J Urol 163:158–162
11. Rabets JC, Jones JS, Patel A et al (2004) Prostate cancer detection with office based saturation biopsy in a repeat biopsy population. J Urol 172:94–97
12. Abouassaly R, Lane BR, Jones JS (2008) Staging saturation biopsy in patients with prostate cancer on active surveillance protocol. Urology 71:573–577
13. Barzell WEWW (2003) Transperineal template guided saturation biopsy of the prostate: rationale, indications, and technique. Urol Times 31:2
14. Onik G, Barzell W (2008) Transperineal 3D mapping biopsy of the prostate: an essential tool in selecting patients for focal prostate cancer therapy. Urol Oncol 26:506–510
15. Sivaraman A, Sanchez-Salas R, Barret E et al (2015) Transperineal template-guided mapping biopsy of the prostate. Int J Urol 22:146–151
16. Sonn GA, Chang E, Natarajan S et al (2014) Value of targeted prostate biopsy using magnetic resonance-ultrasound fusion in men with prior negative biopsy and elevated prostate-specific antigen. Eur Urol 65:809–815
17. Siddiqui MM, Rais-Bahrami S, Turkbey B et al (2015) Comparison of MR/ultrasound fusion-guided biopsy with ultrasound-guided biopsy for the diagnosis of prostate cancer. JAMA 313:390–397
18. Kuru TH, Roethke MC, Seidenader J et al (2013) Critical evaluation of magnetic resonance imaging targeted, transrectal ultrasound guided transperineal fusion biopsy for detection of prostate cancer. J Urol 190:1380–1386
19. Hoeks CM, Schouten MG, Bomers JG et al (2012) Three-Tesla magnetic resonance-guided prostate biopsy in men with increased prostate-specific antigen and repeated, negative, random, systematic, transrectal ultrasound biopsies: detection of clinically significant prostate cancers. Eur Urol 62:902–909
20. Vourganti S, Rastinehad A, Yerram NK et al (2012) Multiparametric magnetic resonance imaging and ultrasound fusion biopsy detect prostate cancer in patients with prior negative transrectal ultrasound biopsies. J Urol 188:2152–2157
21. Roethke M, Anastasiadis AG, Lichy M et al (2012) MRI-guided prostate biopsy detects clinically significant cancer: analysis of a cohort of 100 patients after previous negative TRUS biopsy. World J Urol 30:213–218

22. Schoots IG, Roobol MJ, Nieboer D et al (2015) Magnetic resonance imaging-targeted biopsy may enhance the diagnostic accuracy of significant prostate cancer detection compared to standard transrectal ultrasound-guided biopsy: a systematic review and meta-analysis. Eur Urol 68:438–450

23. Algaba F, Epstein JI, Aldape HC et al (1996) Assessment of prostate carcinoma in core needle biopsy—definition of minimal criteria for the diagnosis of cancer in biopsy material. Cancer 78:376–381

24. Baisden BL, Kahane H, Epstein JI (1999) Perineural invasion, mucinous fibroplasia, and glomerulations: diagnostic features of limited cancer on prostate needle biopsy. Am J Surg Pathol 23:918–924

25. Srigley JR (2004) Benign mimickers of prostatic adenocarcinoma. Mod Pathol 17:328–348

26. Humphrey PA (2007) Diagnosis of adenocarcinoma in prostate needle biopsy tissue. J Clin Pathol 60:35–42

27. Epstein JNG (2015) Biopsy interpretation of the prostate. Wolters Kluwer Health, Philadelphia, PA

28. Qian J, Wollan P, Bostwick DG (1997) The extent and multicentricity of high-grade prostatic intraepithelial neoplasia in clinically localized prostatic adenocarcinoma. Hum Pathol 28:143–148

29. Epstein JI, Herawi M (2006) Prostate needle biopsies containing prostatic intraepithelial neoplasia or atypical foci suspicious for carcinoma: implications for patient care. J Urol 175:820–834

30. Netto GJ, Epstein JI (2006) Widespread high-grade prostatic intraepithelial neoplasia on prostatic needle biopsy: a significant likelihood of subsequently diagnosed adenocarcinoma. Am J Surg Pathol 30:1184–1188

31. Akhavan A, Keith JD, Bastacky SI et al (2007) The proportion of cores with high-grade prostatic intraepithelial neoplasia on extended-pattern needle biopsy is significantly associated with prostate cancer on site-directed repeat biopsy. BJU Int 99:765–769

32. Schoenfield L, Jones JS, Zippe CD et al (2007) The incidence of high-grade prostatic intraepithelial neoplasia and atypical glands suspicious for carcinoma on first-time saturation needle biopsy, and the subsequent risk of cancer. BJU Int 99:770–774

33. Merrimen JL, Jones G, Walker D et al (2009) Multifocal high grade prostatic intraepithelial neoplasia is a significant risk factor for prostatic adenocarcinoma. J Urol 182:485–490 discussion 490

34. Al-Hussain TO, Epstein JI (2011) Initial high-grade prostatic intraepithelial neoplasia with carcinoma on subsequent prostate needle biopsy: findings at radical prostatectomy. Am J Surg Pathol 35:1165–1167

35. Tosoian JJ, Alam R, Ball MW et al (2018) Managing high-grade prostatic intraepithelial neoplasia (HGPIN) and atypical glands on prostate biopsy. Nat Rev Urol 15(1):55–66

36. Rhamy RK, Buchanan RD, Spalding MJ (1973) Intraductal carcinoma of the prostate gland. J Urol 109:457–460

37. Guo CC, Epstein JI (2006) Intraductal carcinoma of the prostate on needle biopsy: histologic features and clinical significance. Mod Pathol 19:1528–1535

38. McNeal JE, Yemoto CE (1996) Spread of adenocarcinoma within prostatic ducts and acini. Morphologic and clinical correlations. Am J Surg Pathol 20:802–814

39. Tsuzuki T (2015) Intraductal carcinoma of the prostate: a comprehensive and updated review. Int J Urol 22:140–145

40. Kweldam CF, Wildhagen MF, Steyerberg EW et al (2015) Cribriform growth is highly predictive for postoperative metastasis and disease-specific death in Gleason score 7 prostate cancer. Mod Pathol 28:457–464

41. Kweldam CF, Kummerlin IP, Nieboer D et al (2017) Presence of invasive cribriform or intraductal growth at biopsy outperforms percentage grade 4 in predicting outcome of Gleason score 3 + 4 = 7 prostate cancer. Mod Pathol 30:1126–1132

42. Porter LH, Lawrence MG, Ilic D et al (2017) Systematic review links the prevalence of intraductal carcinoma of the prostate to prostate cancer risk categories. Eur Urol 72:492–495

43. Miyai K, Divatia MK, Shen SS et al (2014) Heterogeneous clinicopathological features of intraductal carcinoma of the prostate: a comparison between "precursor-like" and "regular type" lesions. Int J Clin Exp Pathol 7:2518–2526
44. Epstein JI, Egevad L, Humphrey PA et al (2014) Best practices recommendations in the application of immunohistochemistry in the prostate: report from the International Society of Urologic Pathology consensus conference. Am J Surg Pathol 38:e6–e19
45. Jiang Z, Woda BA, Rock KL et al (2001) P504S: a new molecular marker for the detection of prostate carcinoma. Am J Surg Pathol 25:1397–1404
46. Gaudin PB, Epstein JI (1995) Adenosis of the prostate. Histologic features in needle biopsy specimens. Am J Surg Pathol 19:737–747
47. Beach R, Gown AM, De Peralta-Venturina MN et al (2002) P504S immunohistochemical detection in 405 prostatic specimens including 376 18-gauge needle biopsies. Am J Surg Pathol 26:1588–1596
48. Amin MB, Tamboli P, Varma M et al (1999) Postatrophic hyperplasia of the prostate gland: a detailed analysis of its morphology in needle biopsy specimens. Am J Surg Pathol 23:925–931
49. Brimo F, Epstein JI (2012) Immunohistochemical pitfalls in prostate pathology. Hum Pathol 43:313–324
50. Osunkoya AO, Hansel DE, Sun X et al (2008) Aberrant diffuse expression of p63 in adeno-carcinoma of the prostate on needle biopsy and radical prostatectomy: report of 21 cases. Am J Surg Pathol 32:461–467
51. Giannico GA, Ross HM, Lotan T et al (2013) Aberrant expression of p63 in adenocarcinoma of the prostate: a radical prostatectomy study. Am J Surg Pathol 37:1401–1406
52. Jiang Z, Wu CL, Woda BA et al (2004) Alpha-methylacyl-CoA racemase: a multi-institutional study of a new prostate cancer marker. Histopathology 45:218–225
53. Zhou M, Chinnaiyan AM, Kleer CG et al (2002) Alpha-methylacyl-CoA racemase: a novel tumor marker over-expressed in several human cancers and their precursor lesions. Am J Surg Pathol 26:926–931
54. Yang XJ, Tretiakova MS, Sengupta E et al (2003) Florid basal cell hyperplasia of the prostate: a histological, ultrastructural, and immunohistochemical analysis. Hum Pathol 34:462–470
55. Bailar JC III, Mellinger GT, Gleason DF (1966) Survival rates of patients with prostatic cancer, tumor stage, and differentiation—preliminary report. Cancer Chemother Rep 50:129–136
56. Welch HG, Gorski DH, Albertsen PC (2015) Trends in metastatic breast and prostate cancer—lessons in cancer dynamics. N Engl J Med 373:1685–1687
57. Schroder FH, Hugosson J, Roobol MJ et al (2009) Screening and prostate-cancer mortality in a randomized European study. N Engl J Med 360:1320–1328
58. Epstein JI, Amin MB, Reuter VE et al (2017) Contemporary Gleason grading of prostatic carcinoma: an update with discussion on practical issues to implement the 2014 International Society of Urological Pathology (ISUP) consensus conference on Gleason grading of prostatic carcinoma. Am J Surg Pathol 41:e1–e7
59. Choy B, Pearce SM, Anderson BB et al (2016) Prognostic significance of percentage and architectural types of contemporary Gleason pattern 4 prostate cancer in radical prostatectomy. Am J Surg Pathol 40:1400–1406
60. Pierorazio PM, Walsh PC, Partin AW et al (2013) Prognostic Gleason grade grouping: data based on the modified Gleason scoring system. BJU Int 111:753–760
61. Epstein JI, Zelefsky MJ, Sjoberg DD et al (2016) A contemporary prostate cancer grading system: a validated alternative to the Gleason score. Eur Urol 69:428–435
62. Andriole GL, Crawford ED, Grubb RL 3rd et al (2009) Mortality results from a randomized prostate-cancer screening trial. N Engl J Med 360:1310–1319
63. Cooperberg MR, Broering JM, Carroll PR (2010) Time trends and local variation in primary treatment of localized prostate cancer. J Clin Oncol 28:1117–1123
64. Thorson P, Vollmer RT, Arcangeli C et al (1998) Minimal carcinoma in prostate needle biopsy specimens: diagnostic features and radical prostatectomy follow-up. Mod Pathol 11:543–551
65. Epstein JI (1995) Diagnostic criteria of limited adenocarcinoma of the prostate on needle biopsy. Hum Pathol 26:223–229

66. Iczkowski KA, Bostwick DG (2000) Criteria for biopsy diagnosis of minimal volume prostatic adenocarcinoma: analytic comparison with nondiagnostic but suspicious atypical small acinar proliferation. Arch Pathol Lab Med 124:98–107
67. Epstein JI, Allsbrook WC Jr, Amin MB et al (2005) The 2005 International Society of Urological Pathology (ISUP) consensus conference on Gleason grading of prostatic carcinoma. Am J Surg Pathol 29:1228–1242
68. Epstein JI, Egevad L, Amin MB et al (2016) The 2014 International Society of Urological Pathology (ISUP) consensus conference on Gleason grading of prostatic carcinoma: definition of grading patterns and proposal for a new grading system. Am J Surg Pathol 40:244–252
69. Reis LO, Reinato JA, Silva DC et al (2010) The impact of core biopsy fragmentation in prostate cancer. Int Urol Nephrol 42:965–969
70. Fajardo DA, Epstein JI (2010) Fragmentation of prostatic needle biopsy cores containing adenocarcinoma: the role of specimen submission. BJU Int 105:172–175
71. Gupta C, Ren JZ, Wojno KJ (2004) Individual submission and embedding of prostate biopsies decreases rates of equivocal pathology reports. Urology 63:83–86
72. Yfantis HG LO, Silverberg SG (2002) Prostate core biopsies processing: evaluating current practice. United States and Canadian academy of pathology annual meeting, Chicago, IL, pp. 347–1447
73. Kao J, Upton M, Zhang P et al (2002) Individual prostate biopsy core embedding facilitates maximal tissue representation. J Urol 168:496–499
74. Donaldson IA, Alonzi R, Barratt D et al (2015) Focal therapy: patients, interventions, and outcomes—a report from a consensus meeting. Eur Urol 67:771–777
75. Amin MB, Lin DW, Gore JL et al (2014) The critical role of the pathologist in determining eligibility for active surveillance as a management option in patients with prostate cancer: consensus statement with recommendations supported by the College of American Pathologists, International Society of Urological Pathology, Association of Directors of Anatomic and Surgical Pathology, the New Zealand Society of Pathologists, and the Prostate Cancer Foundation. Arch Pathol Lab Med 138:1387–1405
76. van der Kwast TH, Lopes C, Santonja C et al (2003) Guidelines for processing and reporting of prostatic needle biopsies. J Clin Pathol 56:336–340
77. Van der Kwast T, Bubendorf L, Mazerolles C et al (2013) Guidelines on processing and reporting of prostate biopsies: the 2013 update of the pathology committee of the European Randomized Study of Screening for Prostate Cancer (ERSPC). Virchows Arch 463:367–377
78. Iczkowski KA, Casella G, Seppala RJ et al (2002) Needle core length in sextant biopsy influences prostate cancer detection rate. Urology 59:698–703
79. Srigley JR, Delahunt B, Egevad L et al (2014) Optimising pre-analytical factors affecting quality of prostate biopsies: the case for site specific labelling and single core submission. Pathology 46:579–580
80. Boccon-Gibod L, van der Kwast TH, Montironi R et al (2004) Handling and pathology reporting of prostate biopsies. Eur Urol 46:177–181
81. Obek C, Doganca T, Erdal S et al (2012) Core length in prostate biopsy: size matters. J Urol 187:2051–2055

Chapter 5
Multiparametric MRI and MRI/TRUS Fusion Guided Biopsy for the Diagnosis of Prostate Cancer

Viktoria Schütz, Claudia Kesch, Svenja Dieffenbacher, David Bonekamp, Boris Alexander Hadaschik, Markus Hohenfellner, and Jan Philipp Radtke

Abstract

Purpose of this chapter To demonstrate the timing, benefits, limitations and current controversies of multiparametric magnet resonance imaging (mpMRI) combined with fusion guided biopsy and consider how additional incorporation of multivariable risk stratification might further improve prostate cancer (PC) diagnosis.

Recent findings MpMRI has been shown to add important information to the diagnostic pathway for prostate cancer. Fusion biopsy has also shown advantages in comparison to standard practice for biopsy-naïve men and men with previous biopsy in large prospective studies providing level 1b evidence. Adding upfront multivariable risk stratification followed by or combined with mpMRI diagnostic accuracy can further be improved. Regarding active surveillance (AS), mpMRI in combination with fusion biopsy can support initial candidate selection and may help to monitor disease progression. However, mpMRI and fusion biopsy are not without failure and conflicting data exists to what extend (systematic) biopsies can be omitted.

V. Schütz (✉) · S. Dieffenbacher · M. Hohenfellner
Department of Urology, University Hospital Heidelberg, Heidelberg, Germany
e-mail: viktoria.schuetz@med.uni-heidelberg.de

C. Kesch
Department of Urology, University Hospital Heidelberg, Heidelberg, Germany

The Vancouver Prostate Centre, University of British Columbia, Vancouver, Canada

D. Bonekamp
Department of Radiology, German Cancer Research Center (dkfz), Heidelberg, Germany

B. A. Hadaschik
Department of Urology, University Hospital Essen, Essen, Germany

J. P. Radtke
Department of Urology, University Hospital Heidelberg, Heidelberg, Germany

Department of Radiology, German Cancer Research Center (dkfz), Heidelberg, Germany

© Springer Nature Switzerland AG 2018   87
H. Schatten (ed.), *Molecular & Diagnostic Imaging in Prostate Cancer*,
Advances in Experimental Medicine and Biology 1096,
https://doi.org/10.1007/978-3-319-99286-0_5

Summary The integration of mpMRI into the diagnostic pathway for PC can add important information for further decision making, yet more prospective and randomized data is needed to establish reliable procedure standards after mpMRI acquisition.

Keywords mpMRI fusion guided biopsy · Multiparametric MRI · Prostate biopsy · Prostate cancer · Prostate MRI · Risk calculations

Abbreviations

ADC	Apparent diffusion coefficient
AS	Active surveillance
AUC	Area under the curve
DRE	Digital rectal examination
ERSPC	European randomized study of screening for prostate cancer
GS	Gleason score
mpMRI	Multiparametric magnet resonance imaging
NPV	Negative predictive value
PCa	Prostate cancer
PPV	Positive predictive value
PSA	Prostate specific antigen
ROC	Receiver operating characteristics
SB	Systematic biopsy
sPCa	Significant prostate cancer
TB	targeted mpMRI fusion biopsy
TRUS	Transrectal ultrasound

Key Points: MpMRI is increasingly used in the diagnostic pathway for prostate cancer. However, further discussion on how to best integrate mpMRI and fusion guided biopsy into the diagnostic pathway is required.

- MpMRI significantly outperforms standard 12 core TRUS biopsy for detection of significant prostate cancer. It can therefore be used as an upfront screening test.
- Combining mpMRI and clinical parameters in a multivariable risk model further improves diagnostic accuracy.
- There is not yet enough evidence to recommend for or against a standard biopsy (SB) or repeat biopsy in the case of unsuspicious mpMRI or negative pre-biopsy. Until a standard procedure is established decisions need to be made individually.
- Men under active surveillance benefit from mpMRI for both, initial risk stratification and follow-up.

5.1   Introduction

The goal of an accurate diagnostic pathway for prostate cancer (PCa) is the detection of significant disease and on the other hand avoiding the detection of indolent PCa, which can lead to overtreatment and increased patient morbidity. Standard screening parameters such as prostate specific antigen (PSA) and digital rectal examination (DRE) as well as the standard diagnostic 12-core transrectal ultrasound (TRUS) biopsy do not provide sufficient sensitivity and specificity to meet these goals [1–3].

The implementation of multiparametric magnet resonance imaging (mpMRI) combined with fusion biopsy helps to solve this dilemma and increases diagnostic accuracy. However, while increasingly being used in the clinical routine and already being recommended by several urologic societies there is still room for discussion on how to best integrate mpMRI and fusion biopsy into the diagnostic pathway for PCa [4–6]. According to current literature, this chapter aims to discuss the timing, benefits, limitations and current controversies of mpMRI and fusion biopsy and consider how additional incorporation of multivariable risk stratification might further improve PCa diagnosis.

5.2   Using mpMRI as an Upfront Screening Tool

Recently several studies were conducted to evaluate the use of mpMRI as an upfront screening tool. Especially the prospective, multicentric PROMIS study represents a landmark for the use of mpMRI and fusion biopsy in biopsy naïve men [3]. As a reference test template mapping biopsies were used. With a sensitivity of 93% for the detection of significant prostate cancer (sPCa) and a negative predictive value (NPV) of 89% versus 48% and 74% using 12-core TRUS biopsy, mpMRI is considered a useful tool to select men under suspicion for PCa due to elevated PSA or abnormal DRE [3]. Taking mpMRI into consideration could spare 27% of men from primary biopsy while missing only 7% of sPCa presuming the applied biopsy strategy would yield the same detection rate as a template mapping strategy. Similar results showing the limitations of a standard TRUS biopsy were found by Porpiglia et al., who conducted a trial randomly assigning patients to standard TRUS or mpMRI fusion biopsy and found that the mpMRI based diagnostic pathway had a significantly better performance than the standard way [7]. One study that should especially be mentioned is the PRECISION study which was recently published in the New England Journal of Medicine. A total of 500 men were randomized of which 252 underwent MRI-targeted biopsy while the other 248 men received a standard TRUS biopsy. Not only was there a higher percentage of significant PCa detected in the fusion biopsy group (38% vs. 26%) but at the same time the percentage of men with clinically insignificant cancer was also lower than in the standard-biopsy group [8]. This study highlights the superiority of MRI fusion-biopsy compared to a standard TRUS biopsy.

On the other hand one can argue – as did Matthew Cooperberg – that the overall detection rate of prostate cancer is similar in MRI-biopsy as in TRUS-biopsy. There also remains the question on how to define "clinically significant" prostate cancer which is not consistent among different papers [9]. One further aspect is that one cannot foresee the development of a low-risk carcinoma. By non-diagnosis of so called non "clinically significant" prostate cancer one might miss low-risk carcinoma which can develop into more aggressive disease.

There have been attempts to combine mpMRI with clinical parameters for multivariable risk stratification to further improve diagnostic accuracy. Especially PSA is of importance. Adding PSA density helps to increase the NPV of mpMRI. Data from Distler et al. and Washino et al. support abstaining from biopsy in case of unsuspicious mpMRI and low PSA density (<0.15 ng/mL/mL) [10, 11]. For biopsy naïve men only, Thompson et al. reported an increase in the area under the curve (AUC) in receiver operating characteristics (ROC) analysis from 0.78 to 0.88 by combining PSA, prostate volume and age with PIRADS [12]. Radtke et al. developed a risk model based on PSA, prostate volume, DRE, age and PIRADS with an area under the curve (AUC) of 0.83 for biopsy-naïve men to optimize the prediction of non-invasive sPCa-risk. This model helps to advise for or against a prostate biopsy [13]. In a similar approach van Leeuwen et al. used PSA, prostate volume, DRE, age, previous biopsy and PIRADS for their risk model and found and AUC of 0.88 versus 0.80 without taking mpMRI into account [14]. While the different AUCs of these studies can be explained by including (s)PCa prevalence, mpMRI-parameters and slightly different variables, they all show a significant net benefit of including mpMRI. This demonstrates the importance of mpMRI as a primary screening tool. But it should be kept in mind that all these models are still based on biopsy indications which were based on PSA or DRE deviation compared from standard screening threshold values. In a pilot study evaluating three different screening strategies—subjects with PSA $\geq$ 3 ng/ml + systematic biopsy (SB), subjects with PSA $\geq$ 3 ng/ml + mpMRI + TB and subjects with PSA $\geq$ 1.8 ng/ml— Grenabo Bergdahl et al. found a screening strategy using a lowered PSA cut-off $\geq$1.8 ng/ml in combination with mpMRI and TB to be most accurate in detecting significant cancer while avoiding unnecessary biopsies [15].

The high costs of mpMRI and the financial impact this might have on health care systems are still a major concern regarding mpMRI as a standard screening tool. Faria et al. looked into this question by analyzing data on cost effectiveness derived from the PROMIS study [3, 16]. They showed that a diagnostic pathway using mpMRI first and then up to two MRI-targeted biopsies detects more sPCa per pound spent than a strategy using 12-core TRUS biopsy first (sensitivity = 0.95 vs 0.91) and is cost effective (8350 €/QALY gained]) [16]. Contrary Alberts et al. evaluated a pathway which first calculates the risk of having sPCa by the use of the ERSPC risk calculator 4, based on the fifth European Randomized study of Screening for Prostate Cancer (ERSPC) screening round [17]. They then only perform mpMRI and biopsy in subjects with a risk $\geq$20% [17]. This approach would avoid 65% of mpMRIs or standard TRUS biopsies and therefore save money, but on the other hand 17% of sPC are missed [17]. One other aspect to keep in mind regarding cost efficiency is the money saved by avoiding further treatment. Even though mpMRI and fusion biopsy come with higher costs compared to a standard TRUS biopsy a more accurate diagnosis can also result in an overall more cost-effective strategy as expenses for further treatment such as surgery or radiotherapy can be avoided. Also, if MRI and fusion biopsy are performed at centers with high patient volume costs can be further decreased as a more standardized and routine treatment can reduce costs in the future.

5.3   Incorporation of MPMRI Fusion Guided Biopsy into Risk Modeling for Prostate Cancer

A very interesting aspect which shows the accuracy of the mpMRI is that compared to RP-specimen, mpMRI detects 85–95% of index-lesions and significant PCa (sPCa) [18, 19]. Targeted biopsy (TB), mostly used in a fusion biopsy setting, of suspicious mpMRI-lesions improves the detection of sPCa by 30% [20].

To identify men with sPCa and at the same time avoiding unnecessary biopsies, multivariable risk-based approaches have been introduced [21–23]. A risk calculator based on European Randomized Study of Screening for PC (ERSPC) data was developed to put a number on the risk for sPCa. Roobol et al. demonstrated that 33% of standard biopsies can be omitted in men who are at risk of PCa below 12.5% [23]. However, recent RC do not include MRI data. TB of mpMRI-suspicious lesions alone is a promising strategy to reduce overdetection of insignificant disease, but at the same time MRI-invisible sPCa can be missed [20, 24–26]. In contrast to the approach proclaimed by Alberts et al., Radtke et al. and van Leeuwen et al. therefore added pre-biopsy mpMRI to clinical parameters and developed risk calculators to determine an individual sPCa-risk using a validated biopsy approach combining fusion guided TB and transperineal systematic saturation biopsies (SB) as reference on the one hand and transperineal mapping and TB plus 12-core TRUS on the other hand [13, 14]. Van Leeuwen et al. demonstrated that a model combining age, PSA, DRE, prostate volume, a previous biopsy result and mpMRI PI-RADS Likert score outperforms the model of clinical parameters alone with a discrimination of 0.90 in the Area under the curve of Receiver operating characteristics (ROC) curve analysis [14]. The internal validation was performed using bootstrapping with 1000 iterations on the cohort of 398 men from St. Vincent's clinic, Sydney, Australia [14]. On external validation in 198 men from Royal North Shore Private Hospital, Sydney, Australia, the discrimination of the full model slightly decreased to an AUC of 0.86 [14]. Beside the model for biopsy-naïve men, Radtke et al. internally validated a risk model combining PSA, prostate volume, DRE, age and mpMRI PI-RADS Likert scoring for men after previous negative biopsy [13]. The model was compared to a validated clinical parameter risk calculator (ERSPC RC 4) and PI-RADS and significantly outperformed both tools alone [13]. Comparing risk models including mpMRI and clinical parameters with risk models that are only based on clinical parameter or PIRADS alone, the accuracy of the decision to perform a biopsy in a patient with the suspicion for sPCa can be improved. In conclusion, risk models that include mpMRI are superior to those risk models not only for men prior to initial biopsy but also for patient after previous negative biopsy [13, 14].

One point that must be stressed is that while the detection of significant prostate cancer can be made more accurate one should also keep in mind that an unsuspicious mpMRI or a low PIRADS-Score cannot be used as an argument against proceeding with the biopsy in case of suspicion for prostate cancer. It is argued that MRI fusion biopsy can reduce the detection of indolent prostate cancer. On the other hand, detecting low-risk prostate cancer can improve patient safety as unnecessary treatment can be avoided and disease monitoring can be made more accurate and reliable when selecting patients for active surveillance.

5.4 Avoiding mpMRI Fusion Biopsy Failure

Even though mpMRI is shown to add important information to the diagnostic pathway for sPCa, mpMRI fusion biopsy can also fail. So far four mechanism for the potential failure of mpMRI fusion biopsy have been identified:, mpMRI invisible cancer, inaccurate sampling, mpMRI reader oversight as well as intralesion Gleason Score (GS) heterogeneity [27]. Muthigi and colleagues showed that in 71% of cases where SB detected sPCa and TB did not, the cancerous finding was within the sextant of the target lesion, confirming the result of Cash et al. who identified inaccurate sampling as one of the main reasons for fusion biopsy failure [28]. Similar, Bryk and colleagues identified a combination of TB and ipsilateral SB as the best strategy to detect sPCa and avoid detection of low risk PCa, comparing TB only, TB and ipsilateral SB and TB and contralateral SB in patients with unilateral mpMRI lesion using TB and both sided SB as reference [29]. The finding from those two studies suggest that inaccurate sampling and intralesion Gleason Score (GS) heterogeneity can be avoided by increasing the number of samples taken from the target area. On the other hand Porpiglia et al. found that two targeted cores placed in the center of the lesion are sufficient to accurately depict the index lesion [30]. More studies on this question are needed. Characterized by a repeatedly found negative predictive value for mpMRI of 63–98% the mpMRI fusion biopsy failure caused by mpMRI invisible cancer can only be solved through additional SB [31, 32]. However, most groups combining TB with 12 core SB did not find a significant benefit for the detection of sPCa by the combination of both methods over TB alone [20, 33, 34]. Contrary to that, Filson et al. found the combined biopsy method to detect significant more sPCa than TB or SB alone [35]. This study support our own results, which demonstrate a significant increase in the detection of sPCa by combining TB and SB, but using a median of 24 SB cores [18]. These controversial results lead to the conclusion that the superiority of sPCa detection in a combined biopsy approach compared to a TB only approach increases with the amount of SB, but with the risk of also finding significantly more low risk diseases. The question of whether to omit SB or not might never get entirely solved and decisions should be made individually to biopsy indications and patient's needs.

One further point regarding the quality and possible reasons for failure of mpMRI fusion biopsy is the technique used to carry out the biopsy. A mpMRI fusion biopsy can be done by "cognitive fusion" – i.e. the urologist focuses during the biopsy on areas suspicious in the MRI. Another more expensive version is to use software – offered by different providers – to directly project the lesion onto the ultrasound image. A third technique and the most expensive one is to perform an in-bore biopsy in an open MRI. On the other hand, using this approach systematic biopsies are more difficult to perform. Also, there have not been many studies comparing these different approaches to mpMRI fusion biopsy [9]. So far, a clear recommendation on the most accurate method to be used for mpMRI fusion biopsy cannot be made.

5.5 MpMRI Fusion Guided Biopsy in Men Requiring a Repeat Biopsy

Men with prior negative biopsy and ongoing suspicion for PCa represent a patient group which needs to be monitored closely. Due to prior sampling overall disease prevalence is reduced compared to a biopsy-naïve population, but those patients presenting with ongoing suspicion for PCa suffer due to limited NPV of 12 core TRUS biopsy. MpMRI has been shown beneficial in various studies to monitor tis patient group and should therefore be recommended in a repeat biopsy setting [4–6]. Most recent studies analyze these patients as a subgroup of a larger cohort, but some works pay special attention to this patient group: equivalent to the PROMIS study Simmons et al. evaluated the diagnostic accuracy of mpMRI in men requiring a repeat prostate biopsy (PICTURE study), though only 31% of men had a previous negative biopsy [36]. When using a mpMRI score of ≥ 3 as a positive test result mpMRI has a sensitivity of 97%, a specificity of 22%, a NPV of 91% and a positive predictive value of 47% [36]. The authors conclude that in 14% of men a repeat biopsy can potentially be avoided at the cost of missing 9% sPCa [36]. Hansen et al. demonstrated a significantly improved area under the curve when combining PI-RADS with PSA density (0.82 vs. 0.85) suggesting to only abstain from repeat biopsy in case of unsuspicious mpMRI and low PSA density [37]. Again, no clear evidence exists upon the question when to safely omit SB. However, Arsov et al. analyzed in a prospective randomized trial setting in-bore TB compared to fusion guided TB plus 12-core TRUS-SB. They showed that additional SB had no significant additional benefit on the detection of sPCa [38]. Contrary to that, recent publications comparing TB alone approaches with 24 or 12 core SB demonstrate that a considerably amount of sPCa is missed by TB only [35, 37].

5.6 MpMRI Fusion Guided Biopsy for Men Under Active Surveillance

Men with PCa eligible for active surveillance (AS) represent another important patient group as an accurate risk classification of potentially insignificant disease is absolutely necessary. To reach this goal mpMRI in combination with fusion biopsy can support initial candidate selection and may help to monitor disease progression. Radtke et al. showed in a cohort of 149 men that initial mpMRI and fusion biopsy before AS result in significant lower rates of subsequent AS qualifications (20% vs. 48%) during a two year follow up compared to men who were selected for AS based on 12-core TRUS biopsy [39]. Supporting these results Henderson et al. demonstrated in a prospective trial that the apparent diffusion coefficient (ADC) is a useful marker when selecting patients for AS as a low ADC value is associated with a shorter time to adverse histology [40]. Several recent studies evaluated mpMRI and fusion biopsy in the context of detecting disease progression. Most of them consistently show that mpMRI predicts the risk of pathological progression very well and that patients with stable mpMRI findings only have a low rate of disease progression [41–44]. Adding clinical parameters to the decision making process also appears to be beneficial when selecting patients for AS. Alberts et al. found in a cohort of 210 men no upgrading at baseline, confirmatory or surveillance biopsy in case of unsuspicious mpMRI and PSA density below 0.15 ng/mL suggesting to reduce follow-up biopsy in these cases [17]. However, there is room for discussion regarding whether or not follow up with fusion biopsy limited to mpMRI-visible targets is sufficient. Meng et al. and Frey et al. both report that on combined SB and TB follow-up mpMRI fusion biopsy TB detects a significant higher amount of upgrading than SB, supporting the idea of omitting SB [42, 45]. On the other hand Tran et al., Ma et al. and Recabal et al. found a relevant proportion of higher grade cancer to be detected by SB only, supporting the need for additional SB [43, 44, 46]. These contradicting results can partly be explained by different study parameters including differences in median TB and SB cores but at the same time they also stress the need for further studies regarding the questions of long-term results, serial mpMRI for replacing repeat biopsies and sufficiency of follow-up biopsies limited to mpMRI targets.

5.7 Conclusion

There are many large studies which show the benefits mpMRI adds to the diagnostic pathway for sPCa not only for biopsy-naïve men but also in a repeat biopsy setting [3, 36]. MpMRI in combination with mpMRI fusion guided biopsy makes the detection of sPCa more accurate. Upfront multivariable risk stratification followed by or combined with mpMRI further improves PCa diagnosis. Risk models can be used to decide whether or not to proceed with the biopsy [12–15, 17]. However, mpMRI and fusion biopsy do not spare failure. The risk of inaccurate sampling and intralesion GS heterogeneity responsible for mpMRI fusion guided biopsy failure can be limited by increasing the number of target cores or sector sampling, the mpMRI negative predictive value of 63–98% however causes a persistent limitation [31, 32].

The choice for or against concurrent SB considerably influences both, the rate of under-detection of sPCa and the rate of over-detection of indolent disease. Study results are still inconsistent, so decisions need to be made based on individual risk adapted patient counselling until a standard procedure has been established.

Acknowledgements None. *Conflicts of interest*: Jan P. Radtke is a consultant for UroNav, Saegeling Medizintechnik, Siemens Healthineers, MedCom and Bender Gruppe. The remaining authors report no potential conflicts of interest.

References

1. Barry MJ (2001) Prostate-specific-antigen testing for early diagnosis of prostate cancer. N Engl J Med 344:1373–1377
2. Loeb S, Bjurlin MA, Nicholson J, Tammela TL, Penson DF, Carter HB, Carroll P, Etzioni R (2014) Overdiagnosis and overtreatment of prostate cancer. Eur Urol 65:1046–1055
3. Ahmed HU, El-Shater Bosaily A, Brown LC, Gabe R, Kaplan R, Parmar MK, Collaco-Moraes Y, Ward K, Hindley RG, Freeman A et al (2017) Diagnostic accuracy of multi parametric MRI and TRUS biopsy in prostate cancer (PROMIS): a paired validating confirmatory study. Lancet 6736:32401–32401
4. Rosenkrantz AB, Verma S, Choyke P, Eberhardt SC, Eggener SE, Gaitonde K, Haider MA, Margolis DJ, Marks LS, Pinto P et al (2016) Prostate magnetic resonance imaging and magnetic resonance imaging targeted biopsy in patients with a prior negative biopsy: a consensus statement by AUA and SAR. J Urol 196:1613–1618
5. Graham J, Kirkbride P, Cann K, Hasler E, Prettyjohns M (2014) Prostate cancer: summary of updated NICE guidance. BMJ 348:f7524
6. Mottet N, Bellmunt J, Bolla M, Briers E, Cumberbatch MG, De Santis M, Fossati N, Gross T, Henry AM, Joniau S et al (2017) EAU-ESTRO-SIOG guidelines on prostate cancer. Part 1: screening, diagnosis, and local treatment with curative intent. Eur Urol 71:618–629
7. Porpiglia F, Manfredi M, Mele F, Cossu M, Bollito E, Veltri A, Cirillo S, Regge D, Faletti R, Passera R et al (2017) Diagnostic pathway with multiparametric magnetic resonance imaging versus standard pathway: results from a randomized prospective study in biopsy-naïve patients with suspected prostate cancer. Eur Urol 72:282–288
8. Veeru Kasivisvanathan, M.R.C.S., Antti S. Rannikko, Ph.D., Marcelo Borghi, M.D., Valeria Panebianco, M.D., Lance A. Mynderse, M.D., Markku H. Vaarala, Ph.D., Alberto Briganti, Ph.D., Lars Budäus, M.D., Giles Hellawell, F.R.C.S.(Urol.), Richard G. Hindley, F.R.C.S.(Urol.), Monique J. Roobol, Ph.D., Scott Eggener, M.D., Maneesh Ghei, F.R.C.S.(Urol.), Arnauld Villers, M.D., Franck Bladou, M.D., Geert M. Villeirs, Ph.D., Jaspal Virdi, F.R.C.S.(Urol.), Silvan Boxler, M.D., Grégoire Robert, Ph.D., Paras B. Singh, F.R.C.S.(Urol.), Wulphert Venderink, M.D., Boris A. Hadaschik, M.D., Alain Ruffion, Ph.D., Jim C. Hu, M.D., Daniel Margolis, M.D., Sébastien Crouzet, Ph.D., Laurence Klotz, M.D., Samir S. Taneja, M.D., Peter Pinto, M.D., Inderbir Gill, M.D., Clare Allen, F.R.C.R., Francesco Giganti, M.D., Alex Freeman, F.R.C.Path., Stephen Morris, Ph.D., Shonit Punwani, F.R.C.R., Norman R. Williams, Ph.D., Chris Brew-Graves, M.Sc., Jonathan Deeks, Ph.D., Yemisi Takwoingi, Ph.D., Mark Emberton, F.R.C.S.(Urol.), and Caroline M. Moore, F.R.C.S.(Urol.) for the PRECISION Study Group Collaborators*(2018) MRI-Targeted or Standard Biopsy for Prostate-Cancer Diagnosis. N Engl J Med 378:1767–1777

9. Cooperberg, Matthew R. (2017) Magnetic Resonance Imaging-targeted Prostate Biopsies: Is the Right Technique the Right Question?. Eur Urol 71:532–533

10. Distler FA, Radtke JP, Bonekamp D, Kesch C, Schlemmer H-P, Wieczorek K, Kirchner M, Pahernik S, Hohenfellner M, Hadaschik BA (2017) The value of PSA density in combination with PI-RADS™ for the accuracy of prostate cancer prediction. J Urol. https://doi.org/10.1016/j.juro.2017.03.130

11. Washino S, Okochi T, Saito K, Konishi T, Hirai M, Kobayashi Y, Miyagawa T (2017) Combination of prostate imaging reporting and data system (PI-RADS) score and prostate-specific antigen (PSA) density predicts biopsy outcome in prostate biopsy naïve patients. BJU Int 119:225–233

12. Thompson JE, van Leeuwen PJ, Moses D, Shnier R, Brenner P, Delprado W, Pulbrook M, Böhm M, Haynes AM, Hayen A et al (2016) The diagnostic performance of multiparametric magnetic resonance imaging to detect significant prostate cancer. J Urol 195:1428–1435

13. Radtke JP, Wiesenfarth M, Kesch C, Freitag MT, Alt CD, Celik K, Distler F, Roth W, Wieczorek K, Stock C et al (2017) Combined clinical parameters and multiparametric magnetic resonance imaging for advanced risk modeling of prostate cancer – patient-tailored risk stratification can reduce unnecessary biopsies. Eur Urol. https://doi.org/10.1016/j.eururo.2017.03.039

14. van Leeuwen PJ, Hayen A, Thompson JE, Moses D, Shnier R, Böhm M, Abuodha M, Haynes A-M, Ting F, Barentsz J et al (2017) A multiparametric magnetic resonance imaging-based risk model to determine the risk of significant prostate cancer prior to biopsy. BJU Int. https://doi.org/10.1111/bju.13814

15. Grenabo Bergdahl A, Wilderäng U, Aus G, Carlsson S, Damber J-E, Frånlund M, Geterud K, Khatami A, Socratous A, Stranne J et al (2016) Role of magnetic resonance imaging in prostate cancer screening: a pilot study within the Göteborg randomised screening trial. Eur Urol 70:566–573

16. Faria R, Soares MO, Spackman E, Ahmed HU, Brown LC, Kaplan R, Emberton M, Sculpher MJ (2017) Optimising the diagnosis of prostate cancer in the era of multiparametric magnetic resonance imaging: a cost-effectiveness analysis based on the prostate MR imaging study (PROMIS). Eur Urol. https://doi.org/10.1016/j.eururo.2017.08.018

17. Alberts AR, Schoots IG, Bokhorst LP, van Leenders GJ, Bangma CH, Roobol MJ (2016) Risk-based patient selection for magnetic resonance imaging-targeted prostate biopsy after negative transrectal ultrasound-guided random biopsy avoids unnecessary magnetic resonance imaging scans. Eur Urol 69:1129–1134

18. Radtke JP, Schwab C, Wolf MB, Freitag MT, Alt CD, Kesch C, Popeneciu IV, Huettenbrink C, Gasch C, Klein T et al (2016) Multiparametric magnetic resonance imaging (MRI) and MRI – transrectal ultrasound fusion biopsy for index tumor detection: correlation with radical prostatectomy specimen. Eur Urol 70:846–853

19. Baco E, Ukimura O, Rud E, Vlatkovic L, Svindland A, Aron M, Palmer S, Matsugasumi T, Marien A, Bernhard J-C et al (2015) Magnetic resonance imaging-transectal ultrasound image-fusion biopsies accurately characterize the index tumor: correlation with step-sectioned radical prostatectomy specimens in 135 patients. Eur Urol 67:787–794

20. Siddiqui MM, Rais-Bahrami S, Turkbey B, George AK, Rothwax J, Shakir N, Okoro C, Raskolnikov D, Parnes HL, Linehan WM et al (2015) Comparison of MR/ultrasound fusion-guided biopsy with ultrasound-guided biopsy for the diagnosis of prostate cancer. JAMA 313:390–397

21. Schröder FH, Hugosson J, Roobol MJ, Tammela TLJ, Zappa M, Nelen V, Kwiatkowski M, Lujan M, Määttänen L, Lilja H et al (2014) Screening and prostate cancer mortality: results of the European randomised study of screening for prostate cancer (ERSPC) at 13 years of follow-up. Lancet 384:2027–2035

22. Roobol MJ, Schröder FH, Hugosson J, Jones JS, Kattan MW, Klein EA, Hamdy F, Neal D, Donovan J, Parekh DJ et al (2012) Importance of prostate volume in the European randomised study of screening for prostate cancer (ERSPC) risk calculators: results from the prostate biopsy collaborative group. World J Urol 30:149–155

23. Roobol MJ, Steyerberg EW, Kranse R, Wolters T, van den Bergh RCN, Bangma CH, Schröder FH (2010) A risk-based strategy improves prostate-specific antigen-driven detection of prostate cancer. Eur Urol 57:79–85

24. Mendhiratta N, Rosenkrantz AB, Meng X, Wysock JS, Fenstermaker M, Huang R, Deng F, Melamed J, Zhou M, Huang WC et al (2015) MRI-ultrasound fusion-targeted prostate biopsy in a consecutive cohort of men with no previous biopsy: reduction of over-detection through improved risk stratification. J Urol 194:1601–1606

25. Meng X, Rosenkrantz AB, Mendhiratta N, Fenstermaker M, Huang R, Wysock JS, Bjurlin M, Marshall S, Deng F-M, Zhou M et al (2016) Relationship between prebiopsy multiparametric magnetic resonance imaging (MRI), biopsy indication, and MRI-ultrasound fusion–targeted prostate biopsy outcomes. Eur Urol 69:512–517

26. Vargas HA, Hötker AM, Goldman DA, Moskowitz CS, Gondo T, Matsumoto K, Ehdaie B, Woo S, Fine SW, Reuter VE et al (2016) Updated prostate imaging reporting and data system (PIRADS v2) recommendations for the detection of clinically significant prostate cancer using multiparametric MRI: critical evaluation using whole-mount pathology as standard of reference. Eur Radiol 26:1606–1612

27. Muthigi A, George AK, Sidana A, Kongnyuy M, Simon R, Moreno V, Merino MJ, Choyke PL, Turkbey B, Wood BJ et al (2017) Missing the mark: prostate cancer upgrading by systematic biopsy over magnetic resonance imaging/transrectal ultrasound fusion biopsy. J Urol 197:327–334

28. Cash H, Günzel K, Maxeiner A, Stephan C, Fischer T, Durmus T, Miller K, Asbach P, Haas M, Kempkensteffen C (2016) Prostate cancer detection on transrectal ultrasonography-guided random biopsy despite negative real-time magnetic resonance imaging/ultrasonography fusion-guided targeted biopsy: reasons for targeted biopsy failure. BJU Int 118:35–43

29. Bryk DJ, Llukani E, Taneja SS, Rosenkrantz AB, Huang WC, Lepor H (2017) The role of ipsilateral and contralateral transrectal ultrasound-guided systematic prostate biopsy in men with unilateral magnetic resonance imaging lesion undergoing magnetic resonance imaging-ultrasound fusion-targeted prostate biopsy. Urology 102:178–182

30. Porpiglia F, De Luca S, Passera R, De Pascale A, Amparore D, Cattaneo G, Checcucci E, De Cillis S, Garrou D, Manfredi M et al (2017) Multiparametric magnetic resonance/ultrasound fusion prostate biopsy: number and spatial distribution of cores for better index tumor detection and characterization. J Urol 198:58–64

31. Fütterer JJ, Briganti A, De Visschere P, Emberton M, Giannarini G, Kirkham A, Taneja SS, Thoeny H, Villeirs G, Villers A (2015) Can clinically significant prostate cancer be detected with multiparametric magnetic resonance imaging? A systematic review of the literature. Eur Urol. https://doi.org/10.1016/j.eururo.2015.01.013

32. Thompson JE, Moses D, Shnier R, Brenner P, Delprado W, Ponsky L, Pulbrook M, Böhm M, Haynes A-M, Hayen A et al (2014) Multiparametric magnetic resonance imaging guided diagnostic biopsy detects significant prostate cancer and could reduce unnecessary biopsies and over detection: a prospective study. J Urol 192:67–74

33. Delongchamps NB, Portalez D, Bruguière E, Rouvière O, Malavaud B, Mozer P, Fiard G, Cornud F (2016) Are magnetic resonance imaging-transrectal ultrasound guided targeted biopsies noninferior to transrectal ultrasound guided systematic biopsies for the detection of prostate cancer? J Urol 196:1069–1075

34. Sonn GA, Natarajan S, Margolis DJA, MacAiran M, Lieu P, Huang J, Dorey FJ, Marks LS (2013) Targeted biopsy in the detection of prostate cancer using an office based magnetic resonance ultrasound fusion device. J Urol 189:86–91

35. Filson CP, Natarajan S, Margolis DJA, Huang J, Lieu P, Dorey FJ, Reiter RE, Marks LS (2016) Prostate cancer detection with magnetic resonance-ultrasound fusion biopsy: The role of systematic and targeted biopsies. Cancer. https://doi.org/10.1002/cncr.29874

36. Simmons LAM, Kanthabalan A, Arya M, Briggs T, Barratt D, Charman SC, Freeman A, Gelister J, Hawkes D, Hu Y et al (2017) The PICTURE study: diagnostic accuracy of multiparametric MRI in men requiring a repeat prostate biopsy. Br J Cancer 116:1159–1165

37. Hansen NL, Kesch C, Barrett T, Koo B, Radtke JP, Bonekamp D, Schlemmer H, Warren AY, Wieczorek K, Hohenfellner M et al (2016) Multicentre evaluation of targeted and systematic biopsies using magnetic resonance and ultrasound image-fusion guided transperineal prostate biopsy in patients with a previous negative biopsy. BJU Int. https://doi.org/10.1111/bju.13711

38. Arsov C, Rabenalt R, Blondin D, Quentin M, Hiester A, Godehardt E, Gabbert HE, Becker N, Antoch G, Albers P et al (2015) Prospective randomized trial comparing magnetic resonance imaging (MRI)-guided in-bore biopsy to MRI-ultrasound fusion and transrectal ultrasound-guided prostate biopsy in patients with prior negative biopsies. Eur Urol 68:713–720

39. Radtke JP, Kuru TH, Bonekamp D, Freitag MT, Wolf MB, Alt CD, Hatiboglu G, Boxler S, Pahernik S, Roth W et al (2016) Further reduction of disqualification rates by additional MRI-targeted biopsy with transperineal saturation biopsy compared with standard 12-core systematic biopsies for the selection of prostate cancer patients for active surveillance. Prostate Cancer Prostatic Dis 19(3):283–291

40. Henderson DR, De Souza NM, Thomas K, Riches SF, Morgan VA, Sohaib SA, Dearnaley DP, Parker CC, Van VANJ, Novara G (2016) Nine-year follow-up for a study of diffusion-weighted magnetic resonance imaging in a prospective prostate cancer active surveillance cohort. Eur Urol 69:1028–1033

41. Nassiri N, Margolis DJ, Natarajan S, Sharma DS, Huang J, Dorey FJ, Marks LS (2017) Targeted biopsy to detect gleason score upgrading during active surveillance for men with low versus intermediate risk prostate cancer. J Urol 197:632–639

42. Frye TP, George AK, Kilchevsky A, Maruf M, Siddiqui MM, Kongnyuy M, Muthigi A, Han H, Parnes HL, Merino M et al (2017) Magnetic resonance imaging-transrectal ultrasound guided fusion biopsy to detect progression in patients with existing lesions on active surveillance for low and intermediate risk prostate cancer. J Urol 197:640–646

43. Recabal P, Assel M, Sjoberg DD, Lee D, Laudone VP, Touijer K, Eastham JA, Vargas HA, Coleman J, Ehdaie B (2016) The efficacy of multiparametric magnetic resonance imaging and magnetic resonance imaging targeted biopsy in risk classification for patients with prostate cancer on active surveillance. J Urol 196:374–381

44. Tran GN, Leapman MS, Nguyen HG, Cowan JE, Shinohara K, Westphalen AC, Carroll PR (2017) Magnetic resonance imaging-ultrasound fusion biopsy during prostate cancer active surveillance. Eur Urol 72:275–281

45. Meng X, Rosenkrantz AB, Mendhiratta N, Fenstermaker M, Huang R, Wysock JS, Bjurlin M, Marshall S, Deng F-M, Zhou M et al (2016) Relationship of pre-biopsy multiparametric mri and biopsy indication with MRI-US fusion-targeted prostate biopsy outcomes. Eur Urol 69:512–517

46. Martin Ma T, Tosoian JJ, Schaeffer EM, If TD, Landis P, Wolf S, Macura KJ, Epstein JI, Mamawala M, Carter HB et al (2017) The role of multiparametric magnetic resonance imaging/ultrasound fusion biopsy in active surveillance. Eur Urol 71:174–180

Chapter 6
Applications of Nanoparticles Probes for Prostate Cancer Imaging and Therapy

Tang Gao, Anyao Bi, Shuiqi Yang, Yi Liu, Xiangqi Kong, and Wenbin Zeng

Abstract Prostate cancer (PCa) is the most common type of cancer in men with high morbidity and mortality. However, the current treatment with drugs often leads to chemotherapy resistance. It is known that the multi-disciplines research on molecular imaging is very helpful for early diagnosing, staging, restaging and precise treatment of PCa. In the past decades, the tumor-specific targeted drugs were developed for the clinic to treat prostate cancer. Among them, the emerging nanotechnology has brought about many exciting novel diagnosis and treatments systems for PCa. Nanotechnology can greatly enhance the treatment activity of PCa and provide novel theranostics platform by utilizing the unique physical/chemical properties, targeting strategy, or by loading with imaging/therapeutic agents. Herein, this chapter focuses on state-of-art advances in imaging and diagnosing PCa with nanomaterials and highlights the approaches used for functionalization of the targeted biomolecules, and in the treatment for various aspects of PCa with multifunctional nanoparticles, nanoplatforms and nanodelivery system.

Keywords Prostate cancer · Molecular imaging · Molecular probe · Biomarker · Nanoparticles · Cancer treatment

6.1 Introduction

In 1851, Adams first described prostate cancer (PCa) through histological examination [1]. At that time, peopled defined the case of prostate cancer as a rare disease. But to date, prostate cancer becomes the most common type of cancer for males, particularly in the developed countries [2]. Additionally, the incidence of prostate

Tang Gao and Anyao Bi contributed equally to this work.

T. Gao · A. Bi · S. Yang · Y. Liu · X. Kong · W. Zeng (✉)
Xiangya School of Pharmaceutical Sciences, Central South University, Changsha, China

Molecular Imaging Research Center, Central South University, Changsha, China

© Springer Nature Switzerland AG 2018   99
H. Schatten (ed.), *Molecular & Diagnostic Imaging in Prostate Cancer*,
Advances in Experimental Medicine and Biology 1096,
https://doi.org/10.1007/978-3-319-99286-0_6

cancer has kept increasing. In 2012, more than 1.1 million cases were diagnosed with prostate cancer and 307,000 died [3]. To reduce the public health impact of PCa, research has been focused on developing detection and treatment strategies for PCa [4, 5]. CT and MRI technologies are very useful to diagnosis of prostate cancer in clinic. However, to date these technologies haven't been applied for the intraoperative imaging [6]. Fluorescence imaging would be an ideal approach to detect PCa and the image-guided surgery due to its high sensitivity, real-time, noninvasive and high compatibility [7, 8]. On the other hand, the main clinical treatments for PCa include surgery, radiotherapy and chemotherapy [9–11]. Chemotherapy with drugs is the primary clinical treatment to prolong patient survival [12]. Unfortunately, the serious toxicity of chemotherapeutics to normal tissues, poor penetration into deeper tumor tissues and the chemotherapy resistance limited their efficacy [13, 14]. Recently, the emerging nanotechnology has brought about many exciting novel diagnosis and treatment systems for PCa. Utilizing the unique physical/chemical properties and targeting strategies, or loading with imaging/therapeutic agents, nanotechnology can greatly enhance the treatment activity of PCa and provide a diagnosis/theranostics platform for cancer. Nanotechnology can greatly enhance the treatment activity of PCa and provide novel theranostics platform by utilizing the unique physical/chemical properties, targeting strategy, or by loading with imaging/therapeutic agents. Nanotechnology is promising to diagnosis and treatment of cancer by using the unique properties of engineered nanoparticles [15]. We can benefit a lot from nanotechnology, such as delivery of poorly water-soluble drugs, improving the targeting of drugs, increasing of cell permeability, construction of innovative therapeutic and diagnostic probes [16]. In this review, we focus on the bench-to-bed advances of nanotechnologies for fluorescence diagnosis and treatment of PCa.

6.2 Nanotechnologies for Fluorescence Diagnosis of PCa *in vivo*

Diagnostic can offer phenotype, and stage of cancer and aid in guiding treatment. The multi-disciplines research on molecular imaging is helpful for the early diagnosing, staging, restaging and precise treatment of PCa. Nanotechnology used in diagnostic provides imaging with high sensitivity, resolution, specificity, and reliability. With the developing of nanotechnology and imaging technology, we can detect cancer biomarkers at the molecular and evaluate therapeutic outcomes in vivo. Although CT and MRI are commonly used to diagnosis of prostate cancer in clinic, this section we will focus on fluorescence imaging, due to the high sensitivity, real-time, noninvasive and high compatibility. To realize early detection and imaging in current therapies PCa, various biomarkers of PCa have been discovered, such as prostate specific antigen (PSA), [17, 18] prostate specific membrane antigen (PSMA), [19] hepsin [20] and matriptase [21].

6.2.1   Targeted PSA Nanoprobe for Imaging PCa

PSA is produced by the prostate gland that is a 33 kDa androgen-regulated serine protease. Nowadays, diagnosis of PCa is often relied on the usage of biomarkers, especially PSA. It has been applied as an organ-specific biomarker for a long time and has been one of the most commonly diagnosis index for PCa, leading to the obvious enhanced detection at earlier stage and helping to decrease the number of metastatic patients.

In 2001 Lövgren reported a detection technology based on a europium (III) nanoparticles and successfully demonstrated the concentration detection and visualization of PSA molecules by a time-resolved microscope [22]. They first washed and activated the commercially available europium chelate (β-diketone)-incorporated polystyrene nanoparticles by phosphate buffer and Fluka. Then, 15 mmol/L streptavidin was added into the activated nanoparticles buffer for 2 h incubation. Finally, the particles were loaded with streptavidin. Biotinylated PSA was incubated with streptavidin-coated 107-nm nanoparticles, with a small volume of 30 μL in order to make PSA direct react on the bottom of the plate for detection. The detection limit was 0.38 ng/L, or 10 fmol/L of PSA molecules correspondence. The nanoparticle loaded with streptavidin was more than ten-fold sensitive than the previously reported molecule probe in a microtiter plate-based PSA assay [23]. In addition, the nanoparticles could achieve an obvious visible in a 45-s exposure time to PSA, indicating a good future in clinical application.

In 2006, Lee designed a hybrid probe with artificial tag molecules by combining particle and peptides, which have high specificity to PSA [24]. After the digestion reaction by with PSA, the peptide was cleaved, leading an individual surface enhanced Raman scattering (SERS) of nanoparticles signal change. The probe could achieve PSA proteolytic reactions imaging. The probe was prepared starting from evaporating nanoscale Au layer on polystyrene nanoparticles. Meanwhile, peptides were preparation by the PSA specific substrate sequence (HSSKLQ) and were ended by a Raman tag molecule. The peptides were linked through a Au-S bond of the nanocrescents to the Au surface at last. During the peptide digestion experiments, the peptide-conjugated nanoparticles were incubated with PSA molecules for 2 h on a 37 °C thermal plate. Monitored by the SERS spectra on the peptide digestion experiments, the peaks of the Raman tag molecules, such as 525 cm^{-1} from biotin almost disappeared completely after the digestion reaction had finished. The results indicated that such peptide-conjugated nanoparticles could be applied as a specific probe on the concentration of the cancer biomarker PSA image.

Fluorescent probes with multiplexing capability and improved brightness are in great need for low abundance targets analysis in bioassays and clinical cases. QDs are found to be 20–50 times brighter than single dye molecules, and were of vital importance to various applications owing to their desirable optical properties. Gao developed a new strategy of nanoparticles probe design, successfully demonstrated a sensitive detection of human prostate specific antigen (PSA) probe in 2009 [25]. They developed a new method for preparation of QD based on nanoparticle-amphiphilic polymer complexes self-assembly in homogeneous solution. QDs coupled with polymaleic anhydride-octadecene via multivalent hydrophobic interactions are highly soluble in tetrahydrofuran but form aggregates in polar solvents. As a new approach of the formation mechanism, a great deal of QDs can be loaded into a nanocore and the embedded nanoparticles space distribution could be manipulated. In 2016, Chen reported the application of novel sub-5 nm $Lu_6O_5F_8$:Eu^{3+} nanoprobe for the successful detection of PSA in clinical cases [26]. They have developed inorganic lanthanide fluoride nanoparticles based on dissolution-enhanced luminescent bioassay technique, leading to amplified signal and improving the detection sensitivity. They synthesized monodisperse and ultra-small Ln^{3+} doped lutetium oxyfluoride nanoparticles via a modified thermalde composition route. Ln^{3+}-NPs were activated with EDC and NHS. Then, the activated NPs were purified by centrifuging at 13,600 rpm and incubated with avidin in phosphate buffered saline. Biotinylated anti-PSA monoclonal antibody was added to each well and the plate was incubated. Thereafter, avidin-conjugated $Lu_6O_5F_8$:Eu^{3+} NPs was added to each well and the plate was incubated. The buffer was measured at room temperature under the kinetic and time-resolved detection mode on a multi-modal microplate reader. The limit of detection for PSA was as low as 0.52 pg/mL, almost a 200-fold sensitivity to that of a commercial DELFIA kit, which indicated a highly promising for the early diagnosis of PCa.

6.2.2   Targeted Prostate-Specific Membrane Antigen (PSMA) Nanoprobes for Imaging PCa

The PSMA is expressed in both the benign, and the neoplastic prostatic epithelial cells, and in other tissues, such as kidney, liver and brain. It is up-regulation in metastatic disease and in hormone-resistant states. It is a transmembrane with 750 amino acid and type II glycoprotein which is primarily expressed in normal human prostate epithelium while overexpressed in PCa cells. PSMA is a very significantly target for PCa imaging and therapy because it is expressed by virtually all PCa cells and its expression is further increased in poorly differentiated, metastatic and hormone-refractory carcinomas [27].

Research work indicated that biotinylated anti-PSMA antibody conjugated to streptavidin-labeled iron oxide nanoparticles would be used as the unique probe for detection and diagnosis of PCa cells. In 2013 Berkman exhibited the first AuNPs system for targeting PSMA expressing level in PCa cells with conjugation of a small molecule peptidomimetic inhibitor [28]. The construction of the PSMA-targeted AuNPs was generated by commercially available 5 nm AuNPs coated with streptavidin and incubating the biotinylated PSMA inhibitor. The PSMA-targeted AuNPs was generated by commercially available 5 nm, and the AuNPs was coated with streptavidin and incubated by the biotinylated PSMA inhibitor. After centrifugal filtration to remove redundant biotinylated PSMA inhibitor, the PSMA-targeted nanoparticles has been composed in suspension and characterized by transmission electron microscopy. The PSMA inhibitor-mediated binding test indicated that the PSMA-targeted nanoparticles have a superior significant binding ability to LNCaP cells, compared to non-targeted AuNPs nanoparticles. The results suggested that the unique targeting of PSMA-targeted AuNPs is better than over non-targeted non-specificity AuNPs, and for the first time it demonstrated that AuNPs can be used to target PSMA by the employment of small molecule inhibitors.

6.3 Nanotechnologies for Prostate Cancer Treatment

6.3.1 Treatment of Prostate Cancer via Chemotherapy with Nanomaterials

In clinical practices, current treatments of prostate cancer are predominantly systemically administered chemotherapy, surgery and radiotherapy [29]. Chemotherapy with drugs, such as paclitaxel (PTX), doxorubicin and docetaxel (DTX) is effective to prolong survival and improve quality of life for patients. However, chemotherapeutics can cause many side-effects, such as body weight, hair loss, nausea, cardiac, liver and kidney toxicity and a destructive "bystander" effect to neighboring cells [30, 31]. In addition, due to the poor penetration of drugs into tumor tissues, the therapeutic efficacy is limited [32]. In order to overcome the systemic toxicity and low therapeutic efficacy, many technologies, such as drug analogs, prodrugs and nanomaterials, have been developed for clinical applications [33, 34]. In recent year, nanomaterial has been one of the most promising tools to significantly enhance antitumor efficacy because of their unique intrinsic physical and chemical properties, [35] and more and more studies were devoted to the treatment of prostate cancer via chemotherapy with nanomaterials to increase drug efficacy, decrease drug toxicity, and maintain a relatively high concentration of drug at the site of interest.

As known, poly(D,L-lactic-co-glycolic acid) (PLGA) is an excellent controlled release polymer because of their safety in clinic. In 2008, Farokhzad's group reported a unique nanotechnology to deliver cisplatin to prostate cancer cells [36]. In their strategy, platinum (IV) compound c,t,c-$[Pt(NH_3)_2(O_2CCH_2CH_2CH_2CH_2CH_3)_2Cl_2]$, as a cisplatin-prodrug, was encapsulated in nanoparticles to deliver cisplatin, and the prostate-specific membrane antigen targeting aptamers (Apt) was introduced to decorate the surface of the nanoparticles and target to the prostate cancer cell. The nanoparticles were derived from PEG-functionalized PLGA and used as a controlled release polymer system to deliver and release drugs to target cells with high safety and low clearance. Through the intrastrand cross-links, the cisplatin could be reductive released from the nanoparticles forms. Cell experiments demonstrated that the curative effect of aptamer-derivatized Pt(IV)-encapsulated nanoparticles was better than cisplatin or nontargeted nanoparticles significantly. The *in vivo* result demonstrated that system was efficacious in reducing prostate tumors at a significantly low dose of platinum [37]. Further, they codelivered cisplatin and docetaxel to prostate cancer cells through a self-assembled polymeric nanoparticle platform in 2010 [38]. The self-assembled polymeric NPs could target to PSMA through the A10 aptamer on the surface with an outstanding efficacy on PCa. In addition, since NPs size could affect the penetration and distribution of tumor cells through the enhanced permeability and retention effect, more and more studies focused on the size to enhance the drugs to tumor sites and improve the efficacy. For example, C. Furman group designed and synthesized a paclitaxel-loaded small PLGA NPs [39]. The size of NPs was between in 45 and 95 nm. Their results showed that the small paclitaxel-loaded PLGA NPs have better efficacy than the free drug and larger NPs. Besides the PLGA and related materials mentioned above, there are many other materials, such as carboxymethylcellulose (Cellax) NPs, [40] polyethylene glycol hyperbranched polymers [41] and so on, could be applied as the vehicle to deliver chemical drugs to prostate cancer cell. Recently, magnetic nanoparticles (MNPs) have been attracted more attention due to its advantages such as chemical stability, low toxicity, good biocompatibility. Usually, MNPs refer to the nanomaterials containing cobalt (Co) or iron (Fe) as well as their oxides and alloys. They become superparamagnetic at room temperature when its size is below a critical value. Additionally, MNPs have been treated as promising drug delivery vehicles for therapeutic applications. For instance, Masatoshi' group designed and synthesized a MgNPs-Fe_3O_4 nanoprobe to carry drugs to prostate cancer cell, and founded that the nanoprobe could significantly increase ROS production in prostate cancer cell lines and induce oxidative DNA damage [42]. Compared with the chemical drugs alone, the combination of MgNPs-Fe_3O_4 and a low dose of drug have a superior efficacy on prostate cancer cell *in vitro*. In 2015, Wang and co-workers reported a magnetic nanoparticle clusters (MNCs) loading chemotherapeutic agent of DOX and developed the combination of photothermal therapy (PTT) and chemotherapy for destruction of PC3 cells [43]. Due to the near-infrared property of MNCs, DOX@MNCs could be used as both photothermal mediators and drug vehicles, and could be applied in the combination of PTT and drug delivery for therapy of prostate cancer. The *in vitro* results showed that a higher therapeutic efficacy could be obtained by the chemophotothermal therapy of DOX@MNCs. Recently, gold nanoparticles were also considered as ideal drug delivery platforms due to their nonimmunogenicity and nontoxicity. Moreover, they were synthesized easily, and the high surface area could increase drug density. For example, Liang and co-workers developed a targeted drug delivery strategy based on GSH-stabilized gold NPs (Au@GSH NPs) consisting of a platinum (IV) drug and a receptor targeting peptide CRGDK [44]. Their results indicated that the cytotoxicity and uptake efficiency of this NPs is superior to that of Au@GSH and Au@Pt(IV) systems, and further demonstrated potent cytotoxicity against prostate cancer cells that overexpress Nrp-1 receptors.

In the recent years, some anti-cancer compounds have been confirmed to have the potential to improve effectiveness of current cancer chemotherapies. For example, some natural products such as curcumin [45, 46] epigallocathechin-3-gallate (EGCG), [47] resveratrol taxanes [48] have been encapsulated or loaded in nanoparticles and exhibited significant efficacy against prostate cancer. Moreover, the vascular disruptive agents (VDAs) have been known to synergistically enhance radiation and chemotherapy. Bischof's group designed and synthesized a gold nanoparticle conjugated VDA to significantly improve VDA tumor specific action in combination with locally applied thermal therapy in prostate cancer [49].

6.3.2   Treatment of Prostate Cancer via Gene Delivery with Nanomaterials

As one of the most effective approach in cancer cure, gene delivery has caused wide concern over the recent years. To realize cancer gene therapy, toxic genes need to be diverted to cancer cells and toward cells death steadily and accurately [50]. As a significant regulator for various conditions including developmental, physiological, and pathological, microRNA (miRNA), an endogenously expressed non-coding RNA molecule, have been regarded as potential therapeutic targets in many disease [51, 52]. While, because of the existence of cell membranes and other obstacles, naked genes cannot realize cancer gene therapy alone. Therefore, an adequate vector that can divert the genes efficiently and preserve it from degradation in the blood stream should be designed at once [53]. Recently, non-viral gene delivery systems, including lipids, polymers and nanomaterials, have been developed for siRNA delivery [54, 55]. Frank's group reported the delivery of small interfering RNA (siRNA) through LbL-assembled microcapsules [56]. In his report, based on the LbL(layer-by-layer) assembly of a crosslinked poly(methacrylic acid) film, two different types of microcapsules were used to deliver an siRNA targeting survivin and the expression of the anti-apoptotic protein was observed. The function of this film is to maintain capsule integrity in the oxidizing bloodstream and in the extracellular environment, thereby, protecting the siRNA from denaturation and make sure the siRNA was released in the reducing intracellular environment. Similarly, Joseph's group reported the fabrication of poly(lactic acid-co-glycolic acid)/siRNA nanoparticles coated with lipids by a unique soft lithography particle molding process named particle replication in nonwetting templates (PRINT) [57]. Combining polymers and lipids, hybrid NPs with high drug encapsulation yields, tunable and sustained drug release profiles, and excellent serum stabilities could makes it applicable drug delivery platform [58].

Polycationic monodispersed poly(L-lysine) (PLL) is a promising carrier among the variety of polymers designed for gene delivery as the result of controllable size, shape, and the feasible for flexible chemical modification [59, 60]. Nevertheless, the relatively low transfection efficiency limited the application of PLL-based polyplexes in clinical treatment [61]. Through PEGylation of poly-L-lysine-cholic acid (PLL-CA), a kind of amphiphilic polycations have been synthesized [62]. With 'stealth' capacity, the benzoic imine linker between PEG and PLL-CA is stable at physiological pH. It is cleavable at lower pH especially in the extracellular environment of tumours and the interior of endosomes/lysosomes. It was reported that the solid lipid PEI hybrid nanocarrier has various advantages including the high silencing efficiency in vitro and in vivo, and the low poisonousness and immunogenicity. As one of the most popular polycationic polymers, polyethylenimine (PEI) was widely used as nonviral gene carriers [63]. Because of the high charge density, PEI molecules can form well-condensed complexes with nucleic acids and can strengthen the interaction with cell surfaces [64]. Furthermore, nucleic acids can be released efficiently from the endosomes through proton sponge effect [65]. Those outstanding properties make contribution to the high transfection efficiency of PEI among nonviral gene carriers. Wang's group reported a lipid PEI hybrid nanocarrier (LPN) which combining linear PEI with hydrophobic, hexadecyl groups (hydrophobic hexadecylated polyethylenimine (H-PEI)) [66]. The LPN would solved or improved several key issues of siRNA/PEI systems. It includes physical encapsulation of the siRNA rather than coating them on carrier surface, reduction of the loss of siRNA and easiness of controlled, continuing intracellular siRNA release, prevented cells from quick exposure to a high level of unencapsulated PEI molecules, provided more sites for grafting cell-targeting [67–70]. While, the severe cytotoxicity of PEI caused by the high density of positive charge was discovered and limited the application of PEI [71]. Contrapose this phenomenon, a kind of non-viral cationic polymer vector mPEG-PEI nanoparticles was used as a carrier and the shRNA plasmid was rebuilt [72]. With the engrafted of moieties polyethylene glycol, PEI polymers showed a lower cytotoxicity and better stability. To further increase cell biocompatibility, disulfide linkage was introduced in the branched PEI (SSPEI) containing multiple amine backbone [73]. SSPEI polymer labeled with poly-arginine (R11) which has the highest uptake by different prostate cancer cell lines compared with other four cell permeable peptide was used to deliver miR-145 to the prostate cancer. Moreover, SSPEI polymer introduced a polyethylene glycol chain linker which could enhance biocompatibility and extend circulation time in the bloodstream [74, 75]. The result showed that the R11-SSPEI/FAM-miR-145 complex could dramatically inhibit tumor growth and prolong survival time. To build a better gene delivery system, the ability of target is significant.

With the development of gene therapy technology, therapeutic effects of single gene-targeted therapy was regard as limited, and multiple gene silencing was proposed. Recently, the combinatorial RNAi technology and simultaneous multiple gene silencing have been attempted to cancer therapy and received a big success [76–79]. Therefore, a new class of dual-genes targeted two different sequences of siRNA (vascular endothelial growth factor (VEGF) and B-cell lymphoma 2 (Bcl-2)) and its their delivery systems for efficient cancer treatment was were developed [80]. Carrying glycol chitosan nanoparticles, the dual-poly-siRNA encapsulated thiolated glycol chitosan (tGC) nanoparticles (dual-NPs) can provided efficient and controlled dual-poly-siRNA delivery and achieved multi-gene silencing with synergistic effects of cancer therapy. Recently, researches showed that the suppression of crucial gene products such as REV1, REV3L can resistant the sensibility of tumors to chemotherapy reduce the drug resistance of relapsed tumors during the error-prone translation DNA synthesis pathway. Based on those researches, a promising strategy which combining siRNA based therapeutics with traditional DNA-damaging chemotherapy was proposed for treating patients with malignancies [81, 82]. A versatile nanoparticle platform was developed to deliver REV1/REV3L-specific siRNAs and a cisplatin prodrug to the same tumor cells simultaneously. Obviously, the result showed a better therapeutic efficacy both in vitro and in vivo than signal single cisplatin prodrug or REV1/REV3L-specific siRNAs [83]. To overcome the accumulation of chemotherapeutic agent in tumor tissue, a synergistic and selective inhibition of cancer cell proliferation platform was reported [84]. With high positive charges on the surface, the DTX-encapsulated bovine serum albumin-polyethylenimine layer-by-layer (LBL) nanoparticles (DTX/BSA-PEILBL NPs) can could adsorb the negative charged p44/p42 MAPK siRNA efficiently. And then, branched polyethylenimine (bPEI) was adsorbed on the surface to form DTX/BSA-PEILBL/siRNA NPs. The result reported a less values of IC_{50} and a higher median survival, provided a promising synergistic delivery system for clinical treatment of PCa. To realize the application of RNAi therapeutic regimen in clinical treatment of prostate cancer, an approach to evaluate the siRNA delivery at the intended site of action is significant. Therefore, theranostics nanoparticles that associated imaging with therapeutic features was proposed and developed [85]. A nice platform for theranostic imaging of prostate cancer was designed and developed [86]. This theranostic nanoparticle was combined by three core components including the prodrug-activating enzyme bacterial cytosine deaminase (βCD), the imaging carrier poly-L-lysine which traced with a near-infrared fluorescent probe Cy5.5 and the carrier which is not only for siRNA delivery but also for labeling with [^{111}In]DOTA for SPECT imaging. Their results verified the feasibility of the platform for associate detection and treatment. Later, a multimodal theranostic lipid-nanoparticle was reported. The probe was constitutive by a near-infrared (NIR) fluorescent core, covered by phospholipid monolayer, instituted with siRNA payloads with ultra-small particle size (<30 nm) [87]. The siRNA delivery with the orthotopic tumor model was evaluated by image co-registration of computed tomography and fluorescence molecular tomography, achieving efficacious RNAi therapy.

6.3.3 Treatment of Prostate Cancer via Cancer Immunotherapy with Nanomaterials

Cancer immunotherapy is an approach of triggering lymphocyte reaction of cancer related antigen [88]. With the development of tumor-specific therapies, treatments such as peptide-TAAs, protein-TAAs, or cell-based vaccination approaches, was reported and they were potentially capable of stimulating pre-existing antitumor immunity or of inducing de novo antigenic responses. However, after decades of intensive pursuit, this remains a challenging goal. Classical vaccination approaches have been extensively tested and found to be largely inefficient [89, 90]. Whereas, current vaccine design paradigms can effectively generate prophylactic and therapeutic immunities against foreign pathogens, they maybe ill-suited as platforms with which to build cancer fighting vaccines. Compared to conventional approaches, nanoparticles can protect the payload (antigen/adjuvant) from the surrounding biological milieu, increase its half-life, minimize its systemic toxicity, promote its delivery to APCs, or even directly trigger the activation of TAA-specific T-cells. The application of nanomedicine in cancer immunotherapy is currently one of the most challenging areas in cancer therapeutic intervention. Development of nanovaccine formulations was mainly from two directions, nanoparticles as vehicles for drug delivery and nanoparticle-based approaches to elicit antitumor immunity. During the last two decades, several nanoparticle-based compounds delivering encapsulated or conjugated cytotoxic drugs had reached the clinical trial stage [91, 92]. On the other hand, nanoparticle-based delivery TAAs to professional APCs were reported as a potential nanovaccine formulation. It has been shown that certain nanoparticle designs possess immunostimulatory properties, and that antigens delivered by these nanoparticle types can induce T- and B-cell responses in the absence of exogenously added adjuvants [93, 94]. Efficient and targeted delivery of immunomodulatory and immunostimulatory molecules to appropriate cells is vital to the successful development of nanovaccine formulations [95].

Recently, nanoparticles were used as vehicles for drug delivery. For example, Lee had established an interesting platform for effective chemoimmunotherapy. He described a delivery system based on a dendrimer and a single-strand DNA-A9 PSMA RNA aptamer hybrid, and was designed to overcome the drawbacks of conventional cancer therapies. Employing these vehicles, they researchers had demonstrated the promising possibility of this chemoimmunotherapeutic system against prostate cancer both in *in vitro* and *in vivo* models. The system has many advantages including cancer-targeting ability, immune-stimulating function, and drug delivery for chemotherapy. The drug-loaded conjugate showed excellent antitumor efficacy and target specificity in an in vivo prostate tumor model due to the high drug-loading capacity and enhanced stability of oligonucleotides *in vivo*. This proof-of-concept demonstrates the potential value of this nanostructure system (the high drug-loading capacity and enhanced in vivo stability) as a new combination approach for improving cancer treatments [96]. Sun designed a redox-responsive immunostimulatory polymeric prodrug carrier, PSSN10, for programmable co-delivery of an immune checkpoint inhibitor NLG919 (NLG) and a chemotherapeutic doxorubicin. In his work, the prodrug carrier could achieve synergistic therapeutic efficacy, prevent cancer relapse, and combined chemotherapy with immunotherapy as a new modality for tumor treatment. NLG-containing PSSN10 prodrug polymers were self-assembled into nano-sized micelles that served as a carrier to load DOX (DOX/PSSN10 micelles). The PSSN10 carrier dose-dependently enhanced T-cell immune responses in the lymphocyte-Panc02 co-culture experiments, and significantly inhibited tumor growth in vivo. DOX/PSSN10 micelles showed potent cytotoxicity *in vitro* against 4T1.2 mouse breast cancer cells and PC-3 human prostate cancer cells comparable to that of DOX [97]. Successful treatment requires delivery of critical amounts of drug into the cancerous tissue [98–100]. As a model, Jankun and coworkers used LnCAP human prostate cancer cells targeted by antibody (against prostate-specific membrane antigen) to conjugate with hematoporphyrin (HP) through protein-based nanotechnology. Their results suggested that mAb/HP conjugates could deliver HP to the tumor cells and then result in considerably less HP in the circulation and, therefore, lower the delivery of HP to normal tissue, and fewer side effects [101]. Nanoparticle-based approaches can elicit antitumor immunity. Regulating molecular interactions in the T-cell synapse to prevent autoimmunity or, conversely, to boost anti-tumor immunity has long been a goal in immunotherapy. However, delivering therapeutically meaningful doses of immune-modulating compounds into the synapse is still a major challenge [102]. For this purpose, Stephan and coworkers reported a male imide-functionlized nanoparticles by covalent coupling to free thiol groups on T-cell membrane proteins. It could efficient delivery of compounds into the T-cell synapse. They had demonstrated that surface-linked NPs are rapidly polarized toward the nascent immunological synapse (IS) at the T-cell/APC contact zone during antigen recognition. Combination of NSC-87877-the loaded NPs on the surface of tumor specific T cells can cause the tumor site to produce a large number of T cell proliferations before cancer cells to adoptive transfer in mice. Relative to the other of the same drug intake system, nanoparticle-based can improve survival rate of the treated animals [103].

6.4   Conclusion and Future Trends

Molecular imaging probes represent an important, growing class of chemical compound for biology, pharmaceutical sciences, preclinical and clinic studies and further application. In conjunction with the nanoparticle, the identification of molecular imaging targets and the development of new labeled molecular probes for those targets are crucial for expanding the capability of in vivo molecular imaging for biological research, molecular diagnostics and drug discovery.

Various nanomaterials, such as PLGA NPs, cellax NPs, MNPs, AuNPs and etc., have been developed owing to their unique properties. Compared with traditional chemotherapy, the nanomaterial drug/anti-cancer compound delivery system has a better targeting and a lower toxicity, and thus they could exhibit a high therapeutic efficacy for prostate cancer. There is still a long way to realize the application of RNAi therapeutic regimen in clinical treatment of prostate cancer. Too much work need to be done, such as novel miRNA and siRNA with higher efficiency to kill cancer cells, better siRNA delivery system with no cytotoxicity, low accumulation in in tumor tissue, better targeting and other properties that can improve therapeutic efficiency and make patients feel more comfortable, better evaluating and monitoring system toward all aspects of RNAi therapeutic regimen. Combination therapy may be another practicable strategy to get a better therapeutic efficiency. Nanoparticle-based tumor immunotherapy is still in its infancy, but apparently this is a method with great prospects for development. Some researchers have reported various nanoparticles as vehicles for drug delivery to tumor antigens and immune stimulating molecules to DCs and other professional APC type. Compared to conventional approaches, nanoparticles can protect the payload (antigen/adjuvant) from the surrounding biological milieu, increase its half-life, and minimize its systemic toxicity. Similarly, nanoparticle-based approaches aimed at regulating molecular interactions in the T-cell synapse to prevent autoimmunity or, conversely, to boost anti-tumor immunity have also provided preliminary evidence of efficacy.

The design, synthesis and application of dual- and multi-modality probes will be a hot research area, which may be the next generation of probes. The combination of different functional modality undoubtedly will improve the accuracy of diagnosis and analysis to prostate cancer. On the other hand, a targeted gene-therapy approach is also being developed to activate the immune system to recognize prostate cancer cells. To discovery nanoprobes based on labeled gene and related macromolecule and these types of approaches might provide a new direction of prostate cancer therapies. We believe that such imaging probes will play a vital role in further understanding of prostate cancer, for PCa's early detection and more effective treatment.

Acknowledgement We are grateful for the financial supports from National Natural Science Foundation of China (81741134, 81671756 and 81271634), and Key Research Project of Science and Technology Foundation of Hunan Province (2017SK2093).

References

1. Adams J (1853) The case of scirrhous of the prostate gland with corresponding affliction of the lymphatic glands in the lumbar region and in the pelvis. Lancet 1(1):393–393
2. Ito K (2014) Prostate cancer in Asian men. Nat Rev Urol 11(4):197–212
3. Siegel RL, Miller KD, Jemal A (2015) Cancer statistics, 2015. CA Cancer J Clin 65(1):5–29
4. Thompson IM Jr, Cabang AB, Wargovich MJ (2014) Future directions in the prevention of prostate cancer. Nat Rev Clin Oncol 11(1):49–60
5. Ravindranathan P, Lee T-K, Yang L, Centenera MM, Butler L, Tilley WD, Hsieh J-T, Ahn J-M, Raj GV (2013) Peptidomimetic targeting of critical androgen receptor-coregulator interactions in prostate cancer. Nat Commun 4:1923–1934
6. Picchio M, Mapelli P, Panebianco V, Castellucci P, Incerti E, Briganti A, Gandaglia G, Kirienko M, Barchetti F, Nanni C (2015) Imaging biomarkers in prostate cancer: role of PET/CT and MRI. Eur J Nucl Med Mol Imaging 42(4):644–655
7. Nguyen QT, Tsien RY (2013) Fluorescence-guided surgery with live molecular navigation—a new cutting edge. Nat Rev Cancer 13(9):653–662
8. Komljenovic D, Wiessler M, Waldeck W, Ehemann V, Pipkorn R, Schrenk H-H, Debus J, Braun K (2016) NIR-cyanine dye linker: a promising candidate for isochoric fluorescence imaging in molecular cancer diagnostics and therapy monitoring. Theranostics 6(1):131–142
9. Klotz L, Emberton M (2014) Management of low risk prostate cancer [mdash] active surveillance and focal therapy. Nat Rev Clin Oncol 11(6):324–334
10. Mohiuddin JJ, Baker BR, Chen RC (2015) Radiotherapy for high-risk prostate cancer. Nat Rev Urol 12(3):145–154
11. Yasufuku T, Arakawa S, Fujisawa M, Shigemura K, Matsumoto O (2010) Combination chemotherapy with weekly paclitaxel or docetaxel, carboplatin, and estramustine for hormone-refractory prostate cancer. J Infect Chemother 16(3):200–205

12. Tannock IF, de Wit R, Berry WR, Horti J, Pluzanska A, Chi KN, Oudard S, Théodore C, James ND, Turesson I (2004) Docetaxel plus prednisone or mitoxantrone plus prednisone for advanced prostate cancer. N Engl J Med 351(15):1502–1512

13. Sugahara KN, Teesalu T, Karmali PP, Kotamraju VR, Agemy L, Greenwald DR, Ruoslahti E (2010) Coadministration of a tumor-penetrating peptide enhances the efficacy of cancer drugs. Science 328(5981):1031–1035

14. Hambley TW (2009) Is anticancer drug development heading in the right direction? Cancer Res 69(4):1259–1262

15. Kim BY, Rutka JT, Chan WC (2010) Nanomedicine. N Engl J Med 363(25):2434–2443

16. Ferrari M (2005) Cancer nanotechnology: opportunities and challenges. Nat Rev Cancer 5(3):161–171

17. Constantinou J, Feneley MR (2006) PSA testing: an evolving relationship with prostate cancer screening. Prostate Cancer Prostatic Dis 9(1):6–13

18. Esfahani M, Ataei N, Panjehpour M (2015) Biomarkers for evaluation of prostate cancer prognosis. Asian Pac J Cancer Prev 16(7):2601–2611

19. Chang SS (2004) Overview of prostate-specific membrane antigen. Rev Urol 6(Suppl 10):S13

20. Kelly KA, Setlur SR, Ross R, Anbazhagan R, Waterman P, Rubin MA, Weissleder R (2008) Detection of early prostate cancer using a hepsin-targeted imaging agent. Cancer Res 68(7):2286–2291

21. Saleem M, Adhami VM, Zhong W, Longley BJ, Lin C-Y, Dickson RB, Reagan-Shaw S, Jarrard DF, Mukhtar H (2006) A novel biomarker for staging human prostate adenocarcinoma: overexpression of matriptase with concomitant loss of its inhibitor, hepatocyte growth factor activator inhibitor-1. Cancer Epidemiol Biomarkers Prev 15(2):217–227

22. Härmä H, Soukka T, Lövgren T (2001) Europium nanoparticles and time-resolved fluorescence for ultrasensitive detection of prostate-specific antigen. Clin Chem 47(3):561–568

23. Ferguson RA, Yu H, Kalyvas M, Zammit S, Diamandis EP (1996) Ultrasensitive detection of prostate-specific antigen by a time-resolved immunofluorometric assay and the immulite immunochemiluminescent third-generation assay: potential applications in prostate and breast cancers. Clin Chem 42(5):675–684

24. Liu GL, Chen FF, Ellman JA, Lee LP (2006) Peptide-nanoparticle hybrid Sers probe for dynamic detection of active cancer biomarker enzymes. Conf Proc IEEE Engl Med Biol Soc 1:795–798

25. Gao XH (2009) QD barcodes for biosensing and detection. Annu Int Conf IEEE Eng Med Biol Soc 2009:6372–6373

26. Xu J, Zhou S, Tu D, Zheng W, Huang P, Li R, Chen Z, Huang M, Chen X (2016) Sub-5 nm lanthanide-doped lutetium oxyfluoride nanoprobes for ultrasensitive detection of prostate specific antigen. Chem Sci 7(4):2572–2578

27. O'Keefe DS, Bacich DJ, Heston WD (2004) Comparative analysis of prostate-specific membrane antigen (PSMA) versus a prostate-specific membrane antigen-like gene. Prostate 58(2):200–210

28. Kasten BB, Liu T, Nedrowbyers JR, Benny PD, Berkman CE (2013) Targeting prostate cancer cells with PSMA inhibitor-guided gold nanoparticles. Bioorg Med Chem Lett 23(2):565–568

29. Heidenreich A, Bellmunt J, Bolla M, Joniau S, Mason M, Matveev V (2011) Eau guidelines on prostate cancer. Part 1: screening, diagnosis, and treatment of clinically localised disease. Eur Urol 59(1):61–71

30. Yazdan MS, Naghmeh N, Oshani D, Tan A, Seifalian AM (2011) A new era of cancer treatment: carbon nanotubes as drug delivery tools. Int J Nanomedicine 6(1):2963–2980

31. Alexiou C, Schmid RJ, Jurgons R, Kremer M, Wanner G, Bergemann C (2006) Targeting cancer cells: magnetic nanoparticles as drug carriers. Eur Biophys J 35(5):446–450

32. Jia J, Zhu F, Ma X, Cao ZW, Li YX, Chen YZ (2009) Mechanisms of drug combinations: interaction and network perspectives. Nat Rev Drug Discov 8(2):111–128

33. Cheng L, Wang C, Feng L, Yang K, Liu Z (2014) Functional nanomaterials for phototherapies of cancer. Chin J Clin Oncol 114(21):10869–10939

34. Ryu JH, Koo H, Sun IC, Yuk SH, Choi K, Kim K (2012) Tumor-targeting multi-functional nanoparticles for theragnosis: new paradigm for cancer therapy. Adv Drug Deliv Rev 64(13):1447–1458
35. Shapira A, Livney YD, Broxterman HJ, Assaraf YG (2011) Nanomedicine for targeted cancer therapy: towards the overcoming of drug resistance. Drug Resist Updat 14(3):150–163
36. Dhar S, Gu FX, Langer R, Farokhzad OC, Lippard SJ (2009) Targeted delivery of cisplatin to prostate cancer cells by aptamer functionalized pt(iv) prodrug-plga-peg nanoparticles. Proc Natl Acad Sci U S A 2009(45):157–158
37. Dhar S, Kolishetti N, Lippard SJ, Farokhzad OC (2011) Targeted delivery of a cisplatin pro-drug for safer and more effective prostate cancer therapy in vivo. Proc Natl Acad Sci U S A 108(5):1850–1885
38. Kolishetti N, Dhar S, Valencia PM, Lin LQ, Karnik R, Lippard SJ (2010) Engineering of self-assembled nanoparticle platform for precisely controlled combination drug therapy. Proc Natl Acad Sci U S A 107(42):17939–17944
39. Broc-Ryckewaert DL, Carpentier R, Lipka E, Daher S, Vaccher C, Betbeder D (2013) Development of innovative paclitaxel-loaded small plga nanoparticles: study of their anti-proliferative activity and their molecular interactions on prostatic cancer cells. Int J Pharm 454(2):712–719
40. Hoang B, Ernsting MJ, Murakami M, Undzys E, Li S (2014) Docetaxel–carboxymethylcel-lulose nanoparticles display enhanced anti-tumor activity in murine models of castration-resistant prostate cancer. Int J Pharm 471(1–2):224–233
41. Pearce AK, Simpson JD, Fletcher NL, Houston ZH, Fuchs AV, Russell PJ (2017) Localised delivery of doxorubicin to prostate cancer cells through a PSMA-targeted hyperbranched polymer theranostic. Biomaterials 141(1):330–339
42. Sato A, Itcho N, Ishiguro H, Okamoto D, Kobayashi N, Kawai K (2013) Magnetic nanopar-ticles of Fe_3O_4 enhance docetaxel-induced prostate cancer cell death. Int J Nanomedicine 8:3151–3160
43. Zhang W, Zheng X, Shen S, Wang X (2015) Doxorubicin-loaded magnetic nanoparticle clus-ters for chemo-photothermal treatment of the prostate cancer cell line pc3. Biochem Biophys Res Commun 466(2):278–282
44. Kumar A, Huo S, Zhang X, Liu J, Tan A, Li S, Jin S, Xue X, Zhao Y, Ji T, Han L, Liu H, Zhang X, Zhang J, Zou G, Wang T, Tang S, Liang XJ (2014) Neuropilin-1-targeted gold nanoparticles enhance therapeutic efficacy of platinum(iv) drug for prostate cancer treatment. ACS Nano 8(5):4205–4220
45. Yallapu MM, Khan S, Maher DM, Ebeling MC, Sundram V, Chauhan N (2014) Anti-cancer activity of curcumin loaded nanoparticles in prostate cancer. Biomaterials 35(30):8635–8648
46. Yallapu MM, Dobberpuhl MR, Maher DM, Jaggi M, Chauhan SC (2012) Design of curcumin loaded cellulose nanoparticles for prostate cancer. Curr Drug Metab 13(1):120–128
47. Sanna V, Singh CK, Jashari R, Adhami VM, Chamcheu JC, Rady I (2017) Targeted nanopar-ticles encapsulating (−)-epigallocatechin-3-gallate for prostate cancer prevention and ther-apy. Sci Rep 7:41573–41588
48. Narayanan NK, Nargi D, Randolph C, Narayanan BA (2009) Liposome encapsulation of curcumin and resveratrol in combination reduces prostate cancer incidence in pten knockout mice. Int J Cancer 125(1):1–8
49. Shenoi MM, Iltis I, Choi J, Koonce NA, Metzger GJ, Griffin RJ (2013) Nanoparticle deliv-ered vascular disrupting agents (vdas): use of tnf-alpha conjugated gold nanoparticles for multimodal cancer therapy. Mol Pharm 10(5):1683–1694
50. Soltani F, Sankianm HA, Ramezani M (2013) Development of a novel histone H1- based recombinant fusion peptide for targeted non-viral gene delivery. Int J Pharm 441(1–2):307–315
51. Barbato C, Ruberti F, Cogoni C (2009) Searching for MIND: microRNAs in neurodegenera-tive diseases. J Biomed Biotechnol 2009(1):871313–871321

52. Hwang DW, Son S, Jang J, Youn H, Lee S, Lee D (2011) A brain-targeted rabies virus glycoprotein-disulfide linked PEI nanocarrier for delivery of neurogenic microrna. Biomaterials 32(21):4968–4975
53. Li L, Wei Y, Gong C (2015) Polymeric nanocarriers for non-viral gene delivery. J Biomed Nanotechnol 11(5):739–770
54. Park TG, Ji HJ, Kim SW (2006) Current status of polymeric gene delivery systems. Adv Drug Deliv Rev 58(4):467–486
55. Jing GJ, Fu ZG, Dan B, Lin LR, Yang TC, Shi SL (2010) Development and evaluation of a novel nano-scale vector for sirna. J Cell Biochem 111(4):881–888
56. Becker AL, Orlotti NI, Folini M, Cavalieri F, Zelikin AN, Johnston AP, Zaffaroni N, Caruso F (2011) Redox-active polymer microcapsules for the delivery of a survivin-specific sirna in prostate cancer cells. ACS Nano 5(2):1335–1344
57. Hasan W, Chu K, Gullapalli A, Dunn SS, Enlow EM, Luft JC, Tian S, Napier ME, Pohlhaus PD, Rolland JP, Desimone JM (2012) Delivery of multiple sirnas using lipid-coated plga nanoparticles for treatment of prostate cancer. Nano Lett 12(1):287–292
58. De MI, Imbertie L, Rieumajou V, Major M, Kravtzoff R, Betbeder D (2000) Proofs of the structure of lipid coated nanoparticles (smbv) used as drug carriers. Pharm Res 17(7):817–824
59. Walsh M, Tangney M, O'Neill MJ, Larkin JO, Soden DM, Mckenna SL, Darcy R, O'Sullivan GC, O'Driscoll CM (2006) Evaluation of cellular uptake and gene transfer efficiency of pegylated poly-L-lysine compacted DNA: implications for cancer gene therapy. Mol Pharm 3(6):644–653
60. Watanabe K, Harada-Shiba M, Suzuki A, Gokuden R, Kurihara R, Sugao Y, Mori T, Katayama Y, Niidome T (2009) In vivo siRNA delivery with dendritic poly(Llysine) for the treatment of hypercholesterolemia. Mol BioSyst 5(11):1306–1310
61. Guo J, Bourre L, Soden DM, O'Sullivan GC, O'Driscoll C (2011) Can non-viral technologies knockdown the barriers to siRNA delivery and achieve the next generation of cancer therapeutics. Biotechnol Adv 29(4):402–417
62. Guo J, Cheng WP, Gu J, Ding C, Qu X, Yang Z, Yang Z, O'Driscoll C (2012) Systemic delivery of therapeutic small interfering rna using a PH-triggered amphiphilic poly-l-lysine nanocarrier to suppress prostate cancer growth in mice. Eur J Pharm Sci 45(5):521–532
63. Jere D, Jiang HL, Arote R, Kim YK, Choi YJ, Cho MH, Akaike T, Cho CS (2009) Degradable polyethylenimines as DNA and small-interfering RNA carriers. Expert Opin Drug Deliv 6(8):827–834
64. Demeneix B, Behr JP (2005) Polyethylenimine (PEI). Adv Genet 53(1):217–230
65. Dehshahri A, Oskuee RK, Shier WT, Hatefi A, Ramezani M (2009) Gene transfer efficiency of high primary amine content, hydrophobic, alkyl-oligoamine derivatives of polyethylenimine. Biomaterials 30(25):4187–4194
66. Xue HY, Narvikar M, Zhao JB, Wong HL (2013) Lipid encapsulation of cationic polymers in hybrid nanocarriers reduces their non-specific toxicity to breast epithelial cells. Pharm Res 30(2):572–583
67. Xu Z, Chen L, Gu W, Gao Y, Lin L, Zhang Z, Xi Y, Li Y (2009) The performance of docetaxel-loaded solid lipid nanoparticles targeted to hepatocellular carcinoma. Biomaterials 30(2):226–232
68. Pozo-Rodríguez AD, Pujals S, Delgado D, Solinís MA, Gascón AR, Giralt E, Pedraz JL (2009) A proline-rich peptide improves cell transfection of solid lipid nanoparticle-based non-viral vectors. J Control Release 133(1):52–59
69. Wang MT, Jin Y, Yang YX, Zhao CY, Yang HY, Xu XF (2010) In vivo biodistribution, anti-inflammatory, and hepatoprotective effects of liver targeting dexamethasone acetate loaded nanostructured lipid carrier system. Int J Nanomedicine 5(1):487–497
70. Stevens PJ, Sekido M, Lee RJ (2004) A folate-receptor-targeted lipid nanoparticle formulation for a lipophilic paclitaxel prodrug. Pharm Res 21(12):2153–2157

71. Huang W, Lv M, Gao Z (2011) Polyethylenimine grafted with diblock copolymers of poly-ethylene glycol and polycaprolactone as sirna delivery vector. J Control Release 152(Suppl 1):e143–e145
72. Wu Y, Yu J, Liu Y, Yuan L, Yan H, Jing J, Xu G (2014) Delivery of EZH2-shrna with mpeg-PEI nanoparticles for the treatment of prostate cancer in vitro. Int J Mol Med 33(6):1563–1569
73. Son S, Hwang DW, Singha K, Jeong JH, Park TG, Lee DS, Kim WJ (2011) Rvg pep-tide tethered bioreducible polyethylenimine for gene delivery to brain. J Control Release 155(1):18–25
74. Zhang T, Xue X, He D, Hsieh JT (2015) A prostate cancer-targeted polyarginine-disulfide linked PEI nanocarrier for delivery of microRNA. Cancer Lett 365(2):156–165
75. Tarokh Z, Naderi-Manesh H, Nazari M (2016) Towards prostate cancer gene therapy: devel-opment of a chlorotoxin-targeted nanovector for toxic (melittin) gene delivery. Eur J Pharm Sci 99:209–218
76. Tai W, Qin B, Cheng K (2010) Inhibition of breast cancer cell growth and invasiveness by dual silencing of HER-2 and VEGF. Mol Pharm 7(2):543–556
77. Shibata MA, Morimoto J, Shibata E, Otsuki Y (2008) Combination therapy with short inter-fering RNA vectors against VEGF-c and VEGF-α suppresses lymph node and lung metastasis in a mouse immunocompetent mammary cancer model. Cancer Gene Ther 15(12):776–786
78. Han L, Zhang AL, Xu P, Yue X, Yang Y, Wang GX, Jia ZF, Pu PY, Kang CS (2010) Combination gene therapy with PTEN and EGFR siRNA suppresses U251 malignant glioma cell growth in vitro and in vivo. Med Oncol 27(3):843–852
79. Grimm D, Kay MA (2007) Combinatorial RNAi: a winning strategy for the race against evolving targets. Mol Ther 15(5):878–888
80. Lee SJ, Yook S, Yhee JY, Yoon HY, Kim MG, Ku SH, Kim SH, Park JH, Jeong JH, Kwon IC, Lee S, Lee H, Kim K (2015) Co-delivery of VEGF and Bcl-2 dual-targeted siRNA poly-mer using a single nanoparticle for synergistic anti-cancer effects in vivo. J Control Release 220(Pt B):631–641
81. Wang Y, Gao S, Ye WH, Yoon HS, Yang YY (2006) Co-delivery of drugs and DNA from cationic core-shell nanoparticles self-assembled from a biodegradable copolymer. Nat Mater 5(10):791–796
82. Zhang XQ, Xu X, Bertrand N, Pridgen E, Swami A, Farokhzad OC (2012) Interactions of nanomaterials and biological systems: implications to personalized nanomedicine. Adv Drug Deliv Rev 64(13):1363–1384
83. Xu X, Xie K, Zhang XQ, Pridgen EM, Park GY, Cui DS, Shi J, Wu J, Kantoff PW, Lippard SJ, Langer R, Walker GC, Farokhzad OC (2013) Enhancing tumor cell response to chemo-therapy through nanoparticle-mediated codelivery of siRNA and cisplatin prodrug. Proc Natl Acad Sci U S A 110(46):18638–18643
84. Pang ST, Lin FW, Chuang CK, Yang HW (2017) Co-delivery of docetaxel and p44/42 mapk sirna using PSMA antibody-conjugated BSA-PEI layer-by-layer nanoparticles for prostate cancer target therapy. Macromol Biosci 17(5):1600421
85. Tandon P, Farahani K (2011) Nci image guided drug delivery summit. Cancer Res 71(2):314–317
86. Chen Z, Penet MF, Nimmagadda S, Li C, Banerjee SR, Winnard PT, Artemov JD, Glunde K, Pomper MG, Bhujwalla ZM (2012) PSMA-targeted theranostic nanoplex for prostate cancer therapy. ACS Nano 6(9):7752–7762
87. Lin Q, Jin CS, Huang H, Ding L, Zhang Z, Chen J, Zheng G (2014) Nanoparticle-enabled, image-guided treatment planning of target specific RNAi therapeutics in an orthotopic pros-tate cancer model. Small 10(15):3072–3082
88. Shao K, Singha S, Clementecasares X, Tsai S, Yang Y, Santamaria P (2015) Nanoparticle-based immunotherapy for cancer. ACS Nano 9(1):16–30
89. Rosenberg SA, Yang JC, Restifo NP (2004) Cancer immunotherapy: moving beyond current vaccines. Nat Med 10(9):909–915

90. Mocellin S, Mandruzzato S, Bronte V, Lise M, Nitti D (2004) Part I: vaccines for solid tumours. Lancet Oncol 5(11):681–689
91. Cho K, Wang X, Nie S, Chen ZG, Shin DM (2008) Therapeutic nanoparticles for drug delivery in cancer. Clin Cancer Res 14(5):1310–1316
92. Taurin S, Nehoff H, Greish K (2012) Anticancer nanomedicine and tumor vascular permeability; where is the missing link? J Control Release 164(3):265–275
93. Zolnik BS, Asadrieh GF (2010) Nanoparticles and the immune system. Endocrinology 151(2):458–465
94. Dwivedi PD, Tripathi A, Ansari KM, Shanker R, Das M (2011) Impact of nanoparticles on the immune system. J Biomed Nanotechnol 7(1):193–194
95. Leleux J, Roy K (2013) Micro and nanoparticle-based delivery systems for vaccine immunotherapy: an immunological and materials perspective. Adv Healthc Mater 2(1):72–94
96. Lee IH, An S, Yu MK, Kwon HK, Im SH, Jon S (2011) Targeted chemoimmunotherapy using drug-loaded aptamer–dendrimer bioconjugates. J Control Release 155(3):435–441
97. Sun JJ, Chen YC, Huang YX, Zhao WC, Liu YH, Venkataramanan R, Lu BF, Li S (2017) Programmable co-delivery of the immune checkpoint inhibitor NLG919 and chemotherapeutic doxorubicin via a redox-responsive immunostimulatory polymeric prodrug carrier. Acta Pharmacol Sin 38(6):823–834
98. Allison RR, Mota HC, Bagnato VS, Sibata CH (2008) Bio-nanotechnology and photodynamic therapy-state of the art review. Photodiagnosis Photodyn Ther 5(1):19–28
99. Jankun J, Keck RW, Skrzypczak-Jankun E, Lilge L, Selman SH (2005) Diverse optical characteristic of the prostate and light delivery system: implications for computer modelling of prostatic photodynamic therapy. BJU Int 95(9):1237–1244
100. Mitton D, Ackroyd R (2008) A brief overview of photodynamic therapy in Europe. Photodiagnosis Photodyn Ther 5(2):103–111
101. Jankun J (2011) Protein-based nanotechnology: antibody conjugated with photosensitizer in targeted anticancer photoimmunotherapy. Int J Oncol 39(4):949–953
102. Carreño LJ, González PA, Bueno SM, Riedel CA, Kalergis AM (2011) Modulation of the dendritic cell-t-cell synapse to promote pathogen immunity and prevent autoimmunity. Immunotherapy 3(4):6–11
103. Stephan MT, Stephan SB, Bak P, Chen J, Irvine DJ (2012) Synapse-directed delivery of immunomodulators using t-cell-conjugated nanoparticles. Biomaterials 33(23):5776–5787

Chapter 7
Castration-Resistant Prostate Cancer: Mechanisms, Targets and Treatment

André Mansinho, Daniela Macedo, Isabel Fernandes, and Luís Costa

Prostate cancer is the most common malignancy in men, and remains the second leading cause of cancer-related death in this gender [1]. Data suggests that 10–20% of patients with prostate cancer metastasis develop castration-resistant prostate cancer (CRPC) within 5 years of follow-up, and that the median survival since development of castration resistance is approximately 14 months (range 9–30) [2]. Additionally, patients with non-metastatic CRPC are at higher risk of disease progression. Approximately 15–33% of patients develop metastasis within 2 years, increasing the mortality burden in this population [3, 4].

Treatment of metastatic CRPC (mCRPC) is palliative, and disease evolution is often associated with significant morbidity. Before 2010, docetaxel chemotherapy was the only treatment showing a survival advantage, which translated in its approval by the US Food and Drug Administration (FDA), and in its widespread use as first-line therapy globally [5, 6]. More recently, however, several large randomized clinical trials have led to the approval of new agents for the treatment mCRPC. New therapies have all demonstrated an overall survival (OS) benefit in patients with mCRPC who progressed after docetaxel therapy [7]. Also the new generation hormonal manipulations—abiraterone and enzalutamide—have shown an OS benefit in asymptomatic or minimally symptomatic patients who had not received prior chemotherapy [8].

A. Mansinho · D. Macedo
Oncology Department, Hospital de Santa Maria, Centro Hospitalar Lisboa Norte,
Lisbon, Portugal

I. Fernandes · L. Costa (✉)
Department of Oncology, Hospital de Santa Maria, Centro Hospitalar Lisboa Norte,
Lisbon, Portugal

Oncology Division, Faculdade de Medicina de Lisboa, Instituto de Medicina Molecular,
Lisbon, Portugal

© Springer Nature Switzerland AG 2018
H. Schatten (ed.), *Molecular & Diagnostic Imaging in Prostate Cancer*,
Advances in Experimental Medicine and Biology 1096,
https://doi.org/10.1007/978-3-319-99286-0_7

Therapeutic strategies with a symptomatic purpose, such as external radiotherapy, chemotherapy with mitoxantrone or radioisotopes such as samario-153, may also be used. Additionally, the use of bone metabolism-modifying agents, such as denosumab or zoledronate, has shown efficacy in the prevention of skeletal complications in this setting.

7.1   Castration-Resistance

The mainstay of treatment for metastatic hormone-sensitive prostate cancer is androgen deprivation therapy (ADT), aiming at the suppression of circulating testosterone. The goal of ADT is to decrease circulating testosterone to "castrate levels," corresponding to a serum measurement lower than 50 ng/dL. The decline of testosterone to castrate levels results in a decrease in cancer cell proliferation, with subsequent induction of apoptosis. Despite the anti-proliferative response to ADT, cancer cells eventually become resistant to therapy, and signs and/or symptoms of progression are observed in most patients [9, 10]. "Castration resistant" designation is applied to prostate cancer when a measurable progression of disease is observed at the castrate level, detected either by a sequential rise in prostate specific antigen (PSA), or by imaging findings (computed tomography, magnetic resonance imaging, or radionuclide bone scintigraphy). The "castration resistant" designation is privileged over the previously used designations of "androgen independent" and "hormone refractory" disease because, despite absence of circulating testosterone, the tumor remains functionally dependent on androgens and on the androgen receptor [10, 11].

7.2   Treatment of mCRPC

7.2.1   Next Generation Hormonal Therapies

Initial treatment of metastatic prostate cancer consists of androgenic depletion by orchidectomy or luteinizing hormone releasing hormone (LHRH) agonists/antagonists, which may be associated with antiandrogens. Due to the tumor hormonal dependency, LHRH axis blockade should be maintained *ad eternum* in mCRPC, as observed in the SWOG 9346 study.

Testosterone and dihydrotestosterone are the major agonists of the androgen receptor. Leydig cells produce approximately 97% of circulating testosterone, which is converted into dihydrotestosterone in prostate by the 5-alpha-reductase enzyme, the remaining being synthesized in the adrenal gland. When pharmacological or surgical castration is performed, dihydrotestosterone may still be detected in tumor tissues at sufficiently high levels to activate the androgen receptor. Regardless of where it is generated, conversion of dihydrotestosterone precursor through CYP17A1 expression-dependent enzymatic reactions will always be necessary. This was the rational underlying the development of potent CYP17A1 inhibitors [12].

Between 2011 and 2012, new hormonal therapies (abiraterone and enzalutamide) emerged as approved treatments for mCRPC.

Abiraterone is a derivative of pregnenolone, which prevents androgen biosynthesis by inhibiting CYP17A1 at the gonad and extra-gonadal levels and in tumor tissues, leading to an effective androgen depletion [12]. In 2011, the COU-AA-301 Phase 3 study, including 1195 symptomatic mCRPC patients previously treated with docetaxel, compared abiraterone 1000 mg (once daily [qd]) plus prednisone 5 mg (twice daily [bid]) with placebo plus prednisone 5 mg (bid). The study showed an increase in progression-free survival (PFS) (5.6 months vs 3.6 months, $p < 0.001$) and OS (15.8 months vs. 11.2 months, HR = 0.65, $p < 0.001$) with abiraterone [1]. A sub-analysis of the COU-AA-301 study investigated pain control in symptomatic patients post-docetaxel chemotherapy. Results showed that patients in the abiraterone acetate plus prednisone arm experienced more palliation (45% vs 28.8%; $p < 0.001$) and faster median time to palliation of pain (5.6 vs 13.7 months; $p = 0.002$) than those in the placebo arm [13, 14].

Enzalutamide is an androgen receptor inhibitor that blocks several steps of the androgen receptor signaling pathway. It has a high affinity for the ligand domain of the androgen receptor (approximately 5–8 times higher than bicalutamide). The AFFIRM study, in 2012, included 1199 symptomatic mCRPC patients previously treated with taxanes, and compared enzalutamide 160 mg (qd) with placebo. This study found a PFS and OS benefit (8.3 months vs 2.9 months, HR 0.40, $p < 0.001$; 18.4 months vs. 13.6 months, HR 0.63, $p < 0.001$, respectively) associated with enzalutamide (Table 7.1) [15].

More recently, Phase 3 studies evaluated these agents as first-line treatment of asymptomatic or minimally symptomatic mCRPC patients prior to chemotherapy. In 2013, the COU-AA-302 study randomized 1088 patients with no visceral disease to treatment with abiraterone 1000 mg (qd) plus prednisone 5 mg (bid) or placebo plus prednisone 5 mg (bid). Treatment with abiraterone translated in an advantage of PFS (16.5 months vs 8.3 months, HR 0.53, $p < 0.001$) and OS (34.7 months vs 30 days, HR 0.80, $p = 0.0027$) [16]. In 2014, the PREVAIL study recruited 1717

Table 7.1 Efficacy of abiraterone and enzalutamide in the second-line treatment of mCRPC

	Overall survival		
COU-AA-301	Median (months)	Hazard ratio (IC 95%)	p
Abiraterone 1000 mg/dia + prednisolone 5 mg per os bid	15.8	0.74 (0.64–0.86)	<0.0001
Placebo + prednisolone 5 mg per os bid	11.2		
AFFIRM			
Enzalutamide 160 mg/dia	18.4	0.63 (0.53–0.75)	<0.0001
Placebo	13.6		

Table 7.2 Efficacy of abiraterone and enzalutamide in the first-line treatment of mCRPC

| COU-AA-302 | Overall Survival | | |
	Median (months)	Hazard Ratio(IC 95%)	P
Abiraterone 1000 mg/dia + prednisolone 5 mg *per os bid*	34.7	0.81 (0.70–0.93)	0.0033
Placebo + prednisolone 5 mg *per os bid*	30.3		
PREVAIL			
Enzalutamide 160 mg/dia	35.3	0.77 (0.67–0.88)	0.0002
Placebo	31.3		

patients, including those with visceral metastasis, to receive enzalutamide 160 mg or placebo once daily. An OS benefit (35.3 months vs 31.3 months, HR 0.71, $p < 0.001$) was observed in the enzalutamide arm (Table 7.2) [17].

Both abiraterone and enzalutamide are currently approved for the first-line treatment of asymptomatic or minimally symptomatic mCPRC patients, and for the second-line treatment of symptomatic mCPRC patients who failed docetaxel.

These agents are better tolerated than cytostatic therapy. Due to inhibition of CYP17A, abiraterone suppresses the production of androgens and cortisol, with an increase of ACTH. This results in the production of mineralocorticoids, with associated side effects. Hypertension, fluid retention and hypokalaemia are the most common adverse events, although a slight increase in transaminases and a very small percentage of grade 3–4 side effects can also be observed. Supplementation with 5 mg of prednisone (bid) is, therefore, recommended.

Enzalutamide is also a well-tolerated drug. In the AFFIRM and PREVAIL trials, adverse events observed in both study arms consisted of fatigue, diarrhea and facial flushing. As a risk of seizures was reported for some patients in both trials (five out of 800 patients in the AFFIRM trial, and one out of 1717 patients in the PREVAIL trial), a risk/benefit evaluation should be made before starting therapy in patients with a prior history of epilepsy. Although hepatotoxicity has been described as an adverse effect of other antiandrogens, it was not observed in the AFFIRM or PREVAIL trials. The glucocorticoid receptor has been postulated as responsible for enzalutamide resistance in the presence of androgen receptor inhibition, due to overlap with the androgen receptor at various DNA binding sites and to rescue of gene transcription expression previously inhibited by enzalutamide [18]. Therefore, it is recommended that glucocorticoids are discontinued when starting enzalutamide, since there is no need for replacement therapy.

In either indication, therapy should be maintained until disease progression, with the first recommended imaging evaluation performed at 12 weeks, and a total PSA determination performed every month. Progression is assumed:

1. In presence of bone scan with ≥2 lesions, 12 or more weeks after initiation of therapy, confirmed according to PCWG2;
2. in second-line, post-docetaxel therapy of symptomatic patients, when in presence of at least three:

(a) Progression of total PSA 25% above baseline, with a minimum increase of 5.0 ng/mL;
(b) Radiographic progression defined by one of the following:

Bone scan with ≥2 lesions not due to flare effect, confirmed according to PCWG2;
Radiographic evidence of progression of lesions assessed by modified RECIST criteria;

3. clinical or symptomatic progression defined by one of the following:

(a) pain worsening in two consecutive evaluations (>30% increase in bone or visual pain scales or >30% increase with opioid use);
(b) bone events (pathological fracture, spinal cord compression, surgery or radiation to the bone);
(c) need to increase prednisone dose or to switch to a more potent glucocorticoid to treat cancer-related symptoms.

7.2.2 Chemotherapy

The use of cytostatic agents in mCPRC began in the 1990s with the use of mitoxantrone. A randomized Phase 3 study compared mitoxantrone plus corticosteroids to corticosteroids alone, showing a benefit of treatment with mitoxantrone in the control of pain and improvement in quality of life, but not in OS. This paradigm was maintained until 2004, when accumulating evidences supported the use of docetaxel. At this time, two Phase 3 clinical trials were published, establishing the OS benefit associated with the use of docetaxel: the SWOG 99-16 and TAX-327 trials [19, 20].

SWOG 99-16 compared docetaxel 60 mg/m^2 (D2) plus estramustine 280 mg (three times a day [tid];D1–D5) with mitoxantrone 12 mg/m^2 (D1) plus prednisone 5 mg (bid) given every 3 weeks (q3w) in 770 patients with mCPRC. The study evidenced a statistically significant increase in OS in the docetaxel plus estramustine arm (17.5 months vs 15.6 months, HR 0.8 p = 0.01) [19]. TAX-327 compared two dosages of docetaxel (30 mg/m^2 EV weekly for 5 weeks in 6 week cycles and 75 mg/m^2 given q3w plus prednisone 5 mg bid with mitoxantrone 12 mg/m^2 q3w plus prednisone 5 mg bid in 1006 patients with mCRPC. Median OS was 18.9 months in the docetaxel q3w arm, 17.4 months in the docetaxel weekly arm and 16.5 months in the mitoxantrone arm, with only the first group showing a statistically significant advantage (HR 0.79 p = 0.004) [20]. This study led to the approval of docetaxel 75 mg/m^2 q3w plus prednisone 5 mg bid as first-line treatment of mCPRC, due to toxic effects and lack of additional efficacy of the estramustine combination (Table 7.3).

Because TAX-327 and SWOG 99-16 trials allowed a maximum number of 10 and 12 treatment cycles, respectively, the benefit of additional treatment cycles was investigated in a retrospective analysis by Pond G. et al. This analysis included the patient populations of the TAX-327 and the CS-205 trial treatment arms, which compared the administration of docetaxel 75 mg/m^2 q3w plus prednisone 5 mg bid plus

Table 7.3 Docetaxel efficacy data in mCRPC

	Overall survival		
TAX 327	Median (months)	Hazard ratio (IC 95%)	P
Docetaxel 75 mg/m², D1 + prednisolone 5 mg *per os bid* (q3w)	18.9	0.79 (0.67–0.93)	0.004
Docetaxel 75 mg/m² D1 semanal, D1, w1–5 + prednisolone 5 mg *per os bid* (q6w)	17.3	0.86 (0.74–1.02)	0.086
Mitoxantrone 12 mg/m² + prednisolone 5 mg *per os bid* (q3w)	16.5		
SWOG 99-16			
Docetaxel 60 mg/m², D2 + Estramustine 280 mg per os, D1–D5 (q3w)	17.5	0.8 (0.67–0.97)	0.02
Mitoxantrone 12 mg/m² + prednisolone 5 mg *per os bid* (q3w)	15.6		

AT-101 with docetaxel 75 mg/m² q3w plus prednisolone 5 mg bid plus placebo. Although patients completed 17 treatment cycles, there was no survival advantage in completing more than ten consecutive cycles of treatment [21].

According to a study by *Kume H. et al.*, intermittent docetaxel therapy was shown to be feasible in selected patients, based on response assessment. According to the study protocol, therapy should be discontinued if total PSA levels drop below 4 ng/mL, with at least 50% reduction over the target level at treatment start, and should be restarted if total PSA levels rise above over 2 ng/ml, with at least 50% increase over the nadir. Among 51 patients included in the study, 27 (52.9%) were eligible for intermittent therapy. The median interval without therapy was 266 days for the first interruption, and 129.5 days for the second interruption. An OS benefit was observed in the intermittent therapy group (HR 2.98, p = 0.023), probably reflecting a subgroup of patients with a more indolent-, better prognosis-disease, amenable to benefit from this strategy and from its reduced cumulative toxicity [22]. Similar results were observed in a retrospective analysis of the ASCENT trial where, with a similar protocol, PSA response rates higher than 50% were observed in 45.5% of patients after a median of 126 days without therapy [23]. In a recent retrospective analysis by *Oudard S. et al.*, favorable responses to docetaxel therapy were observed (total PSA decrease>50%). Furthermore, retreatment with docetaxel was possible, with no marked difference in OS (18.2 months vs. 16.8 months, p = 0.35) compared to other non-taxane-based therapies, and a progression-free interval longer than 6 months was predictive of better response [24].

For patients with comorbidities, where docetaxel at a 75 mg/m² q3w dose is expected to lead to therapy postponement or discontinuation due to toxicity, the 50 mg/m² every 2 weeks (q2w) regimen may be an option. A reduction in grade 3–4 adverse events, including neutropenia and febrile neutropenia, was observed with the q2w regimen, without compromising efficacy: time to progression was 5.6 months vs 4.9 months (p = 0.014), and OS was 19.5 months vs 17.0 months (p = 0.021) in favor of the q2w regimen [25].

The most recent data on the role of chemotherapy in metastatic prostate cancer came from the CHAARTED and STAMPEDE studies. These two trials suggest an earlier use of docetaxel chemotherapy in castration-sensitive disease, along with hormone therapy, with an important OS advantage in patients with a high-volume disease (visceral disease and/or ≥4 bone lesions). It is, therefore, extremely important to consider an early therapy start in this setting.

Recommendations for treatment with docetaxel are as follows:

- To complete a minimum of six cycles, up to a maximum of ten, if justified by evidence of clinical benefit; perform imaging evaluation until cycle 4 in case of sustained biochemical progression;
- in patients with a total PSA drop below 4 ng/mL and a reduction ≥50% of target level at treatment start, treatment interruption can be considered. In this case, treatment should be resumed if total PSA levels rise above 2 ng/ml, with an increase of at least 50% of the nadir;
- maintaining docetaxel treatment after an initial response with total PSA > 50%, and stable disease with a ≥ 6-month progression-free interval without chemotherapy, can be an option for selected patients not eligible for other, more effective, therapies;
- in patients with significant comorbidities, the 50 mg/m² every 2-week docetaxel schedule can be an option, retaining efficacy with lower toxicity.

In 2010, the new taxane cabazitaxel was approved for the treatment of mCPRC in patients previously treated with docetaxel. This drug was found to retain antitumor activity when used in P-glycoprotein-overexpressed cell lines and in those with tubulin mutations, partially responsible for resistance to docetaxel [26].

A phase 3 study (TROPIC) compared cabazitaxel 25 mg/m² (EV administered q3w) plus prednisone 5 mg (bid) with mitoxantrone 12 mg/m² (EV administered q3w) in 755 patients with mCRPC that had been previously treated with docetaxel chemotherapy. Results showed an OS advantage with cabazitaxel (15.1 months vs 12.7 months, HR 0.70, p < 0.0001), evidencing its benefit in the second-line chemotherapy setting [27]. The use of cabazitaxel is currently reserved for symptomatic patients following docetaxel therapy, and neutropenia prophylaxis with granulocyte stimulation factors should be considered. Standard dosage and setting for cabazitaxel was further evaluated in the FIRSTANA and PROSELICA trials. There was no survival advantage in using cabazitaxel in first-line and the 20 mg/m² dosage was

Table 7.4 Cabazitaxel efficacy data in the treatment of mCRPC

TROPIC	Overall survival		
	Median (months)	Hazard ratio (IC 95%)	P
Cabazitaxel 25 mg/m² D1 + prednisolone 5 mg *per os bid* (q3w)	15.1	0.70 (0.59–0.83)	<0.0001
Mitoxantrone 12 mg/m² + prednisolone 5 mg *per os bid* (q3w)	12.7		

Table 7.5 Taxane-related adverse effects in the treatment of mCRPC

TAX 327		TROPIC	
Adverse event	Frequency (%)	Adverse event	Frequency (%)
Alopecia	65	Neutropenia G3/G4	82
Fatigue	53	Diarrhea	47
Nausea/emesis	42	Fatigue	37
Neutropenia G3/G4	32	Nausea	34
Diarrhea	32	Emesis	23
Onycholysis	30	Asthenia	20
Peripheral neuropathy	30	Constipation	20
Stomatitis	20	Hematuria	17
Peripheral edema	19	Abdominal pain	12
Dysgeusia	18	Dyspnea	12
Anorexia	17	Fever	12
Dyspnea	15	Arthralgia	11
Myalgia	14	Anemia G3/G4	11
Tearing	10	Febrile neutropenia	8
Epistaxis	6	Thrombocytopenia G3/G4	4
Anemia G3/G4	5		
Febrile neutropenia	3		
Thrombocytopenia G3/G4	1		

equivalent to the 25 mg/m^2 dosage, both in first and second line, with a better toxicity profile (Tables 7.4 and 7.5) [28, 29].

7.2.3 Sipuleucel-T

Sipuleucel-T is an immunotherapeutic agent consisting of activated antigen-presenting cells derived from patient's peripheral mononuclear cells (PBMCs), which are subsequently stimulated *in vivo* with a recombinant fusion protein (prostatic antigen, prostatic acid phosphatase and granulocyte stimulating factors), and reinfused into the patient. This agent has been evaluated in several randomized clinical trials. Although none of these trials showed a PFS benefit, a statistically significant OS benefit was observed. The largest trial was the Phase 3 IMPACT study, published in 2010, which demonstrated an OS increase (25.8 months vs 21.7 months, HR 0.78, p = 0.03) compared to placebo in patients with bone or lymph node metastasis and a chemotherapy-free interval ≥3 months. This trial included a highly selected patient population: more than 80% of patients had no previous cytostatic therapy, 75% of patients had a Gleason score ≤7, 53% of patients had no pain complaints, and 43% had low-bone and bone-only disease [30].

Sipuleucel-T has been recently approved by the European Medicines Agency (EMA) for the treatment of mCRPC. However, the procedure requirements, absence of predictive biomarkers of response and associated costs may limit its use.

7.2.4   Radionuclide Therapy: Radium-223

Radium-223 is an α-emitting, bone-seeking calcium mimetic that selectively targets and binds to areas of increased bone turnover in bone metastasis. The drug is administered by intravenous injection at 4-week intervals, up to a total of six injections. The ALSYMPCA trial was a randomized, double blind, Phase 3 study comparing six injections of radium-223 with placebo in men with CRPC and bone-only metastasis who received, were not eligible to receive, or declined docetaxel chemotherapy [31]. Median OS was longer with radium-223 than with placebo (14.9 vs 11.3 months; hazard ratio 0.70, 0.58–0.83; P < 0.001). Subsequent subgroup analysis showed a survival benefit with radium-223, irrespective of previous docetaxel use [32]. In addition, a significant improvement in median time to first symptomatic skeletal event was observed for radium-223 compared to placebo (15.6 vs 9.8 months; hazard ratio 0.66, 0.52–0.83; P < 0.001). Radium-223 was well tolerated and associated with fewer adverse events than placebo. Although the difference was not statistically significant, a higher rate of diarrhea (25% vs 15%) was seen with radium therapy. Other known side effects include nausea, vomiting, peripheral edema, and hematologic abnormalities (anemia, leukopenia, thrombocytopenia, neutropenia). Radium therapy was also associated with a meaningful improvement in quality of life [31].

7.3   How to Choose the First-Line Therapy

Approved molecules with survival benefit in mCRPC			
Mechanism of action	Molecule	Trial	Survival advantage, months
Androgen receptor	Enzalutamide	AFFIRM (post-docetaxel)	18.4 vs 13.6
		PREVAIL (pre-docetaxel)	35.3 vs 31.3
Androgen synthesis inhibition	Abiraterone	COU-AA-301 (post-docetaxel)	15.8 vs 11.2
		COU-AA-302 (pre-docetaxel)	34.7 vs 31.3
Citotoxicity—Microtubule stabilization	Docetaxel	TAX-327 (first-line)	19.2 vs 16.3
	Cabazitaxel	TROPIC (post-docetaxel)	15.1 vs 12.7
Radionuclide—Calcium mimetic	Radium-223	ALSYMPCA (pre/post docetaxel)	14.9 vs 11.3

The choice of first-line therapy and the therapeutic sequencing are not straightforward, due to the number of available therapies and the absence of randomized clinical trials evaluating their sequence. In 2015, a consensus meeting was held for the first time. It was called the St Gallen Advanced Prostate Cancer Consensus Conference—APCCC 2015, and gathered investigators from leading clinical trials and opinion leaders in an effort to answer key clinical questions. Based on the resulting document and in evidences available in the literature, a therapeutic algorithm is proposed.

According to this algorithm, mCPRC patients are initially assigned to one of two groups: eligible or non-eligible for docetaxel cytostatic therapy.

7.4   Patients Non-eligible for Docetaxel Therapy

The following criteria apply for considering a patient ineligible for docetaxel therapy:

- ECOG Performance status (PS) of 3, and most patients with ECOG PS of 2;
- inadequate bone marrow reserve (absolute neutrophil count <1500 cells/mm^3 or platelet count <100,000 cells/mm^3);
- inadequate organ reserve (total bilirubin increase $\geq$1.5$\times$; AST/ALT>3.5 times the upper limit of normal);
- patient refusal to receive chemotherapy.

Although docetaxel use is not contraindicated in elderly patients, caution should be taken when administering the drug in the geriatric population, due to non-prostate cancer-related comorbidities. This evaluation and decision should be made for each patient individually. These patients can be candidates for new-generation hormonal therapy, although trials demonstrating their benefit (COU-AA-302 and PREVAIL) have not been performed on the geriatric population, and no information exists regarding their risk/benefit ratio or quality of life in this setting.

7.5   Patients Eligible for Docetaxel Therapy

Patients eligible for chemotherapy require a previous evaluation for presence or absence of symptoms, disease site and preexisting adverse prognostic factors.

7.5.1   Definition of Asymptomatic/Minimally Symptomatic Patients

The benefit of docetaxel therapy in this patient population may be questionable, considering the drug's toxicity and the potentially absent symptomatic relief, since patients already are asymptomatic or minimally symptomatic. Nevertheless, there is not an unequivocal choice between chemotherapy and new hormonal manipulations, as there are a fraction of patients who are primarily resistant to the latter. The splicing variant of the androgen receptor, AR-V7, was studied by *Antonarakis et al.*, and it seemed to have conferred resistance to both abiraterone and enzalutamide patients [33, 34]. That effect was not seen in taxane treated patients, with the splicing variant emerging as possible biomarker. This concept has been recently validated by *Howard Scher et al.* and could have an important role in the first-line therapy selection for mCRPC patients [35, 36].

There is no evidence of superiority of abiraterone compared to enzalutamide or vice versa, and the use of these drugs should be evaluated in each patient individually, considering the most favorable toxicity profile. Sipuleucel-T can be used in this indication concomitantly with other therapies.

The definition of "minimally symptomatic patient" is not clear in the PCWG2 criteria, but by analyzing inclusion and exclusion criteria in these trials a definition can be reached:

• the COU-AA-302 trial only included patients with an ECOG PS of 0 or 1, and measured symptomatology by the Brief Pain Inventory—Short Form (BPI-SF) score. Patients with a score of 0 or 1 were considered asymptomatic, while those with a score of 2 or 3, minimally symptomatic. Patients with visceral disease were excluded;
• the PREVAIL trial included patients with ECOG 0–2, BPI-SF score <4, no opioid therapy, and visceral disease;
• the IMPACT trial excluded patients with ECOG>2, visceral disease, and bone events (bone fractures, spinal cord compression or radiation therapy/bone surgery). This trial only included asymptomatic patients at the beginning, and subsequently also the inclusion of symptomatic patients (the criteria for this population was not detailed);

Considering this, an asymptomatic or minimally symptomatic population can be characterized as having:

• an ECOG PS of 0–1;
• pain defined according to BPI-SF scale of 0–1 (asymptomatic) or 2–3 (minimally symptomatic), with no need for opioid therapy;
• no previous bone events (long bone-fracture, spinal cord compression or bone radiation therapy/surgery).

7.5.2   Disease Site

Presence of visceral disease in mCRPC is rare, and associated with worse prognosis. According to Phase 3 trials performed in the first-line setting of mCRPC treatment, only TAX-327, SWOG 99-16 and PREVAIL (12% of patients) included patients with visceral disease. Therefore, recommendations on the use of docetaxel or enzalutamide in this subpopulation should be made based on available evidences.

Hepatic metastasis (versus other visceral metastatic sites) are associated with worse prognosis, as observed in the subanalysis of the AFFIRM study and in a subanalysis performed by *Halabi et al.* in the TAX-327 study [37]. Consequently, although a benefit was observed in patients with visceral metastasis treated with enzalutamide in the PREVAIL study, hepatic metastasis should be considered a factor of poor prognosis, and first-line docetaxel therapy, recommended [38].

7.5.3   Adverse Prognostic Factors

Adverse prognostic factors represent a significant risk for rapidly progressive disease, and should be addressed in the asymptomatic or minimally symptomatic population before a treatment decision between docetaxel or new generation hormonal manipulations be made.

A Gleason score >8 in local disease (pre-radiotherapy or surgery) represents a poor prognosis factor for the development of metastasis and mortality. The same is true for a twofold increase in total PSA time in non-metastatic biochemical recurrence. The latter has also been validated as an adverse risk factor in the metastatic setting, but there was no definite cut-off value; in the TAX-327 study, a PSA doubling time value higher than 55 days was established as prognostic for survival [39].

Time to castration resistance, defined as the time from nadir of total PSA under androgen deprivation until confirmed biochemical progression, is a poor prognostic factor and can also be predictive of response to future hormonal manipulations. It does not seem to have an impact on the PSA response to docetaxel or cabazitaxel [40, 41].

Tumor burden should be considered, evaluating the type of metastization (lymph nodes, bone, lymph nodes and bone, visceral). The prognostic implications of this parameter were highlighted in the CHAARTED and STAMPEDE studies, where docetaxel was associated with a survival improvement in patients with hormone-sensitive disease and high tumor burden (defined as four or more bone lesions and/or visceral disease) [42, 43].

When considering a predominantly bone-metastizing disease, bone involvement is translated in bone turnover and lysis parameter alterations, including alkaline phosphatase. In the TAX-327 study, an alkaline phosphatase elevation above the median value was a poor prognostic factor, as it was in the COU-AA-302 study [16]. LDH elevation is a tumor lysis marker, frequently associated with tumor burden. In the COU-AA-302 study, elevation of this parameter was a poor prognostic factor for survival [44]. A low pre-treatment hemoglobin value may reflect spinal cord involvement due to neoplastic infiltration, related to more advanced disease. Also a ratio higher rate of lymphocytes/neutrophils in peripheral blood prior to therapy is, not only a poor prognostic marker, but also an indicator of worse response to either docetaxel or abiraterone.

The final analysis of the COU-AA-301 and COU-AA-302 trials showed that anemia, alkaline phosphatase elevation, ECOG, time from hormone therapy to other therapy and presence of visceral metastases were poor prognostic factors of survival, and presence of four to six of these parameters in the same patient translated into a global survival lower to 6 months [14, 16].

In presence of such adverse prognostic factors, earlier onset of chemotherapy should be considered instead of new generation hormone manipulations. The reason for this is the risk for rapid progression, which can cause the deterioration of patient's overall status and the potential loss for docetaxel therapeutic window.

7.6   Subsequent Therapies

Data from small cohort retrospective studies suggests that clinical activity of docetaxel following abiraterone acetate and enzalutamide is reduced (with response rates lower than those reported in the TAX-327 trial), and that cabazitaxel activity is maintained (with response rates similar to those observed in the TROPIC study). Based on this data alone, there is currently no sufficiently robust information to determine the best sequence of available drugs for first-line treatment.

Sequencing of the two hormonal manipulations agents also seems to have a low efficacy. Several small retrospective series reported lower response rates and median PFS for the use of abiraterone post-enzalutamide and vice versa, compared with trials in the second-line setting (COU-AA-301 and AFFIRM), although higher PSA response rates were observed with the use of enzalutamide in the second line [18].

In absence of robust evidences concerning the sequencing of different agents, all hypothesis are possible for the referred indications. For a patient treated with docetaxel in first line who is clinically asymptomatic/minimally symptomatic and has no poor prognostic factors, docetaxel re-challenge (using the criteria previously described: total PSA response >50% and range free of disease for more than 6 months without therapy) or, preferably, switch to a secondary hormonal manipulation can be considered. If patient is symptomatic or progressing under docetaxel, cabazitaxel therapy should be given. In case of bone-only metastasis patients, use of radium 223 should be considered (currently only indicated in patients who have had at least two previous lines).

For patients who experienced secondary hormonal manipulations, subsequent treatment with docetaxel should be considered, followed by cabazitaxel or radium 223 (for bone-only metastasis patients). Sequential therapy with another secondary hormone-manipulating agent should be left for salvage therapy in highly selected patients who had an excellent prior response to the first therapy.

7.7 Future Perspectives

Defects in homologous repair deficient (HRD) genes, such as BRCA 1/2, ATM, PALB2, RAD51, FANC and CHEK2 are present in about 1/5–1/4 of mCRPC patients [45]. Activity has been seen with poly adenosine disphosphate-ribose polymerase (PARP) inhibitors in the TOPARP trial [46] and to some extent platinum therapy [47].

PTEN loss is also common in mCRPC, activating AKT signalling [45]. Targeting this pathway has shown significant activity in a phase 2 trial and is now being tested in a phase 3 setting [48, 49].

Germline mutations in mismatch-repair genes (MLH1, MSH2, MSH6 and PMS2) are described in a small percentage (0.6%) of men with mCRPC, but recently an hypermutated phenotype was describe by *Pritchard et al.* in mCRPC, with mismatch repair deficiency reaching 12% of mCRPC patients [50]. Pembrolizumab, a monoclonal antibody targeting programmed death 1 (PD-1), was approved by the Food and Drug Administraion (FDA) for cancers with defective mismatch repair [51]. In unselected patients resistant to enzalutamide this agent has considerable activity, warranting further investigation, specially in patients with defective mismatch repair [52].

Prostate specific membrane antigen (PSMA) is highly expressed in mCRPC. PSMA-ligands can be coupled to radionuclides, such as actinium or lutecium (alfa and beta particles, respectively). These molecules were already tested in phase II trials in a heavily treated population, with promising results [53, 54]. A recombinant T-cell engaging bispecific monoclonal antibody (BiTE) directed against PSMA and the CD3 epsilon subunit of the T cell receptor complex, can have potential immunostimulating and antineoplastic activities and is currently being tested [55].

References

1. Siegel RL, Miller KD, Jemal A (2015) Cancer statistics, 2015. CA Cancer J Clin 65:5–29. https://doi.org/10.3322/caac.21254 pmid:25559415

2. Kirby M, Hirst C, Crawford ED (2011) Characterising the castration-resistant prostate cancer population: a systematic review. Int J Clin Pract 65:1180–1192. https://doi.org/10.1111/j.1742-1241.2011.02799.x pmid:21995694
3. Hirst CJ, Cabrera C, Kirby M (2012) Epidemiology of castration resistant prostate cancer: a longitudinal analysis using a UK primary care database. Cancer Epidemiol 36:e349–e353. https://doi.org/10.1016/j.canep.2012.07.012 pmid:22910034
4. Smith MR, Kabbinavar F, Saad F et al (2005) Natural history of rising serum prostate-specific antigen in men with castrate nonmetastatic prostate cancer. J Clin Oncol 23:2918–2925. https://doi.org/10.1200/JCO.2005.01.529 pmid:15860850
5. Tannock IF, de Wit R, Berry WR et al (2004) TAX 327 investigators. Docetaxel plus prednisone or mitoxantrone plus prednisone for advanced prostate cancer. N Engl J Med 351:1502–1512. https://doi.org/10.1056/NEJMoa040720 pmid:15470213
6. Petrylak DP, Tangen CM, Hussain MH et al (2004) Docetaxel and estramustine compared with mitoxantrone and prednisone for advanced refractory prostate cancer. N Engl J Med 351:1513–1520. https://doi.org/10.1056/NEJMoa041318 pmid:15470214
7. Lorente D, Mateo J, Perez-Lopez R, de Bono JS, Attard G (2015) Sequencing of agents in castration-resistant prostate cancer. Lancet Oncol 16:e279–e292. https://doi.org/10.1016/S1470-2045(15)70033-1 pmid:26065613
8. Antonarakis ES, Lu C, Wang H et al (2014) AR-V7 and resistance to enzalutamide and abiraterone in prostate cancer. N Engl J Med 371:1028–1038. https://doi.org/10.1056/NEJMoa1315815 pmid:25184630
9. Karantanos T, Corn PG, Thompson TC (2013) Prostate cancer progression after androgen deprivation therapy: mechanisms of castrate resistance and novel therapeutic approaches. Oncogene 32:5501–5511. https://doi.org/10.1038/onc.2013.206 pmid:23752182
10. Harris WP, Mostaghel EA, Nelson PS, Montgomery B (2009) Androgen deprivation therapy: progress in understanding mechanisms of resistance and optimizing androgen depletion. Nat Clin Pract Urol 6:76–85. https://doi.org/10.1038/ncpuro1296 pmid:19198621
11. Marques RB, Dits NF, Erkens-Schulze S, van Weerden WM, Jenster G (2010) Bypass mechanisms of the androgen receptor pathway in therapy-resistant prostate cancer cell models. PLoS One 5:e13500. https://doi.org/10.1371/journal.pone.0013500 pmid:20976069
12. Stein M et al (2014) Androgen synthesis inhibitors in the treatment of castration-resistant prostate cancer. Asian J Androl 16:387–400
13. de Bono JS, Logothetis CJ, Molina A et al (2011) COU-AA-301 investigators. Abiraterone and increased survival in metastatic prostate cancer. N Engl J Med 364:1995–2005. https://doi.org/10.1056/NEJMoa1014618 pmid:21612468
14. Fizazi K, Scher HI, Molina A et al (2012) COU-AA-301 investigators. Abiraterone acetate for treatment of metastatic castration-resistant prostate cancer: final overall survival analysis of the COU-AA-301 randomised, double-blind, placebo-controlled phase 3 study. Lancet Oncol 13:983–992. https://doi.org/10.1016/S1470-2045(12)70379-0 pmid:22995653
15. Scher HI, Fizazi K, Saad F et al (2012) AFFIRM investigators. Increased survival with enzalutamide in prostate cancer after chemotherapy. N Engl J Med 367:1187–1197. https://doi.org/10.1056/NEJMoa1207506 pmid:22894553
16. Ryan CJ, Smith MR, Fizazi K et al (2014) Final overall survival (OS) analysis of COU-AA-302, a randomized phase 3 study of abiraterone acetate (AA) in metastatic castration-resistant prostate cancer. Ann Oncol 25(Sppl 4):iv255–iv279
17. Beer TM, Armstrong AJ, Rathkopf DE et al (2014) PREVAIL investigators. Enzalutamide in metastatic prostate cancer before chemotherapy. N Engl J Med 371:424–433. https://doi.org/10.1056/NEJMoa1405095 pmid:24881730
18. Lorente D, Mateo J, Perez-Lopez R, de Bono JS, Attard G (2015) Sequencing of agents in castration-resistant prostate cancer. Lancet Oncol 16:e279–e292
19. Petrylak DP, Tangen CM, Hussain MH et al (2004) Docetaxel and estramustine compared with mitoxantrone and prednisone for advanced refractory prostate cancer. N Engl J Med 351:1513–1520

20. Tannock IF, de Wit R, Berry WR et al (2004) Docetaxel plus prednisolone ou mitoxantrone plus prednisolone for advanced prostate cancer. N Engl J Med 351:1502–1512
21. Pond G et al (2012) Evaluating the value of number of cycles of docetaxel and prednisone in men with metastatic castration-resistant prostate cancer. Eur Urol 61(2):363–369
22. Kume H et al (2015) Intermittent docetaxel chemotherapy is feasible for castration-resistant prostate cancer. Mol Clin Oncol 3:303–307
23. Beer T et al (2003) Intermittent chemotherapy in patients with metastatic androgen-independent prostate cancer. Br J Cancer 89(6):968–970
24. Oudard S et al (2015) Docetaxel rechallenge after an initial good response in patients with metastatic castration-resistant prostate cancer. BJU Int 115(5):744–752
25. Kellokumpu-Lehtinen PL et al (2013) 2-weekly versus 3-weekly docetaxel to treat castration-resistant advanced prostate cancer: a randomised, phase 3 trial. Lancet Oncol 14(2):117–124
26. Vrignaud P, Semiond D, Lejeune P et al (2013) Preclinical antitumor activity of cabazitaxel, a semisynthetic taxane active in axaneresistant tumors. Clin Cancer Res 19:2973–2983
27. Oudard S, Fizazi K, Sengeløv L et al (2017) Cabazitaxel versus docetaxel as first-line therapy for patients with metastatic castration-resistant prostate cancer: a randomized phase III trial-FIRSTANA. J Clin Oncol 35(28):3189–3197
28. Eisenberger M, Hardy-Bessard AC, Kim CS et al (2017) Phase III study comparing a reduced dose of cabazitaxel (20 mg/m²) and the currently approved dose (25 mg/m²) in postdocetaxel patients with metastatic castration-resistant prostate cancer-PROSELICA. J Clin Oncol 35(28):3198–3206
29. de Bono JS, Oudard S, Ozguroglu M et al (2010) Prednisone plus cabazitaxel or mitoxantrone for metastatic castration-resistant prostate cancer progressing after docetaxel treatment: a randomized open-label trial. Lancet 376:1147–1154
30. Kantoff P et al (2010) Sipuleucel-T immunotherapy for castration-resistant prostate cancer. N Engl J Med 363:411–422
31. Parker C, Nilsson S, Heinrich D et al (2013) ALSYMPCA Investigators. Alpha emitter radium-223 and survival in metastatic prostate cancer. N Engl J Med 369:213–223. https://doi.org/10.1056/NEJMoa1213755 pmid:23863050
32. Hoskin P, Sartor O, O'Sullivan JM et al (2014) Efficacy and safety of radium-223 dichloride in patients with castration-resistant prostate cancer and symptomatic bone metastases, with or without previous docetaxel use: a prespecified subgroup analysis from the randomised, double-blind, phase 3 ALSYMPCA trial. Lancet Oncol 15:1397–1406. https://doi.org/10.1016/S1470-2045(14)70474-7 pmid:25439694
33. Antonarakis ES, Lu C, Wang H et al (2014) AR-V7 and resistance to enzalutamide and abiraterone in prostate cancer. N Engl J Med 371(11):1028–1038
34. Antonarakis ES, Lu C, Luber B et al (2015) Androgen receptor splice variant 7 and efficacy of taxane chemotherapy in patients with metastatic castration-resistant prostate cancer. JAMA Oncol 1(5):582–591
35. Scher H, Graf RP, Schreiber NA et al (2018) Validation of nuclear-localized AR-V7 on circulating tumor cells (CTC) as a treatment-selection biomarker for managing metastatic castration-resistant prostate cancer (mCRPC). J Clin Oncol 36(Suppl 6S):abstr 273
36. Scher H, Lu D, Schreiber NA et al (2016) Association of AR-V7 on circulating tumor cells as a treatment-specific biomarker with outcomes and survival in castration-resistant prostate cancer. JAMA Oncol 2(11):1441–1449
37. Scher HI, Fizazi SK, Saad F et al (2012) Increased survival with enzalutamide in prostate cancer after chemotherapy. N Engl J Med 367:1187–1197
38. ClinicalTrials.gov (acedido em 25/09/2015). A safety and efficacy study of oral MDV3100 in chemotherapy-naive patients with progressive metastatic prostate cancer (PREVAIL)
39. Armstrong AJ, Garrett-Mayer ES, Yang YC et al (2007) A contemporary prognostic nomogram for men with hormone-refractory metastatic prostate cancer: a TAX327 study analysis. Clin Cancer Res 13(21):6396–6403

40. Loriot Y, Massard C, Albiges L et al (2012) Personalizing treatment in patients with castrate-resistant prostate cancer: a study of predictive factors for secondary endocrine therapies activity. J Clin Oncol 30(Suppl 5):abstr 213
41. Angelergues A, Maillet D, Fléchon A et al (2014) Duration of response to androgen-deprivation therapy and efficacy of secondary hormone therapy, docetaxel, and cabazitaxel in metastatic castration-resistant prostate cancer. J Clin Oncol 32(Suppl 4):abstr 282
42. Sweeney C, Chen YH, Carducci MA et al (2014) Impact on overall survival (OS) with che-mohormonal therapy versus hormonal therapy for hormone-sensitive newly metastatic prostate cancer (mPrCa): an ECOG-led phase III randomized trial. J Clin Oncol 32(5 suppl):abstr LBA2
43. ClinicalTrials.gov (acedido em 25/09/2015) STAMPEDE: systemic therapy in advancing or metastatic prostate cancer: evaluation of drug efficacy: a multi-stage multi-arm randomised controlled trial
44. Fizazi K, Scher HI, Molina A et al (2012) Abiraterone acetate for treatment of metastatic castration-resistant prostate cancer: final overall survival analysis of the COU-AA-301 ran-domised, double-blind, placebo controlled phase 3 study. Lancet Oncol 13:983–992
45. Robinson D, Van Allen EM, Wu YM et al (2015) Integrative clinical genomics of advanced prostate cancer. Cell 161:1215–1228
46. Mateo J, Carreira S, Sandhu S et al (2015) DNA-repair defects and olaparib in metastatic prostate cancer. N Engl J Med 373:1697–1708
47. Cheng HH, Pritchard CC, Boyd T, Nelson PS, Montgomery B (2016) Biallelic inactivation of BRCA2 in platinum-sensitive metastatic castration-resistant prostate cancer. Eur Urol 69:992–995
48. de Bono JS, De Giorgi U, Massard C et al (2016) PTEN loss as a predictive biomarker for the Akt inhibitor ipatasertib combined with abiraterone acetate in patients with metastatic castration-resistant prostate. Ann Oncol 27(Suppl 6):7180 Abstract
49. Ipatasertib plus abiraterone plus prednisone/prednisolone, relative to placebo plus abiraterone plus prednisone/prednisolone in adult male patients with metastatic castrate-resistant prostate cancer (IPATential150). https://clinicaltrials.gov/ct2/show/NCT03072238
50. Pritchard CC, Morrissey C, Kumar A et al (2014) Complex MSH2 and MSH6 mutations in hypermutated microsatellite unstable advanced prostate cancer. Nat Commun 5:4988
51. FDA grants accelerated approval to pembrolizumab for first tissue/site agnostic indication. Food and Drug Administration, Silver Spring, MD, 30 May 2017. https://www.fda.gov/drugs/informationondrugs/approveddrugs/ucm560040.htm
52. Graff JN, Alumkal JJ, Drake CG et al (2016) Early evidence of anti-PD-1 activity in enzalutamide-resistant prostate cancer. Oncotarget 7:52810–52817
53. Rahbar K, Ahmadzadehfar H, Kratochwil C et al (2017) German multicenter study investigating 177Lu-PSMA-617 radioligand therapy in advanced prostate cancer patients. J Nucl Med 58:85–90
54. Kratochwil C, Bruchertseifer F, Rathke H et al (2017) Targeted α-therapy of metastatic castration-resistant prostate cancer with 225Ac-PSMA-617: dosimetry estimate and empiric dose finding. J Nucl Med 58:1624–1631
55. Hernandez-Hoyos G, Sewell T, Bader R et al (2016) MOR209/ES414, a novel bispecific anti-body targeting PSMA for the treatment of metastatic castration-resistant prostate cancer. Mol Cancer Ther 9:2155–2165

Chapter 8
Peptide-Based Radiopharmaceuticals for Molecular Imaging of Prostate Cancer

Tamila J. Stott Reynolds, Charles J. Smith, and Michael R. Lewis

Abstract Given the high incidence of prostate cancer, there is a continuing need for advances in early detection and in effective treatments. Over the last several years, radiolabeled peptides have been developed, which can target receptors on prostate tumors with high affinity and specificity. These peptides are eliminated from normal tissues rapidly, producing high contrast for PET and SPECT imaging. Receptors of interest for tumor imaging include prostate specific membrane antigen (PSMA), gastrin-releasing peptide receptor (GRPR), and $\alpha_v\beta_3$ integrin. Because radiolabeled peptides afford high tumor-to-normal tissue uptake ratios, the potential of peptide-based targeted radiotherapy of prostate cancer is being explored. In addition, targeting either of two receptors with one peptide may allow more tumors to be detected and aid in the delineation of early versus advanced disease. Taken together, all these developments in peptide-based imaging and therapy of prostate cancer offer the promise of personalized, molecular medicine for individual patients.

Keywords Peptides · Radiopharmaceuticals · PET · SPECT · Gastrin-releasing peptide receptor · $\alpha_v\beta_3$ integrin

T. J. Stott Reynolds
Laboratory Animal Resources Center, University of Texas at El Paso, El Paso, TX, USA
e-mail: tjstottreynolds@utep.edu

C. J. Smith
Department of Radiology, University of Missouri, Columbia, MO, USA
e-mail: smithcj@health.missouri.edu

M. R. Lewis (✉)
Department of Veterinary Medicine and Surgery, University of Missouri,
Columbia, MO, USA
e-mail: LewisMic@missouri.edu

© Springer Nature Switzerland AG 2018   135
H. Schatten (ed.), *Molecular & Diagnostic Imaging in Prostate Cancer*,
Advances in Experimental Medicine and Biology 1096,
https://doi.org/10.1007/978-3-319-99286-0_8

8.1 Introduction

Prostate cancer is a significant problem among American men, with one out of seven diagnosed within his lifetime. However, the incidence of prostate cancer-related death (one in every 39 men) is relatively low. Regardless, the poor prognosis and comorbidities of metastatic disease, including excruciating pain, plus the current state of palliative therapy provide motivation to develop innovative new methods for early detection and treatment. In addition, tactics to distinguish indolent from aggressive cancer are critical in improving patient outcome [1–3]. Current therapies, including surgery, external beam radiation, brachytherapy, and chemotherapy, are often unsuccessful in stopping the disease from progressing to a hormone refractory, then inevitably metastatic state. These factors account for most of the complications of and incidence of prostate cancer-related deaths [4, 5]. Moreover, the side effects of the existing therapies are often the result of imprecise delivery. The current diagnostic techniques, such as serum prostate specific antigen (PSA) levels and needle biopsy, are controversial in their ability to achieve accurate, early detection. According to some experts, these methods may produce false positive diagnoses leading to unneeded treatment. However, other experts believe that these methods have demonstrated a beneficial effect on overall mortality [6–13].

Current methods of detection for prostate cancer include the digital rectal exam (DRE), measurement of PSA in the blood, and ultrasonic imaging. If these preliminary diagnostics indicate the possibility of prostate cancer, more advanced imaging and ultrasound-directed biopsy can confirm and help stage the cancer, as well as formulate a prognosis for the patient. The Gleason scoring system based on histological grades is also still key for staging and determining prognoses. For example, a Gleason score ≥7 and a PSA result >20 ng/mL are considered indicators of aggressive behavior and metastasis. The Gleason scoring system has evolved over the decades but remains imperfect in many ways [14–16], and it is possible for two different patients to receive an identical numeric score despite variability in their individual criteria used for scoring. Thus, screening and initial biopsies are a topic of much debate as they are not without risks to the patient's health and wellbeing. The outcome for most diagnoses will be a "watch and wait" or "active monitoring" approach. A great many new cases will not be considered an immediate health crisis for the patient, due to early stage and expected slow progression or the presence of other serious comorbidities that accompany advanced age [17, 18]. Magnetic resonance imaging (MRI), computed tomography (CT), radiopharmaceutical bone scans, and dissection of the pelvic lymph nodes are modalities used to stage prostate cancer further.

Once initial staging is complete, a patient's therapeutic options are determined. Options may only include "watchful waiting/active surveillance" as previously noted through at least Stage III and even to Stage V (depending on patient variables such as age and concurrent health status), as defined by the American Joint Committee on Cancer (AJCC) Tumor, Nodes, Metastasis (TNM) classification for prostate cancer [19]. If a patient elects active intervention, actions may include chemotherapy (*e.g.*, bisphosphonates, docetaxel, cabazitaxel). Other first line treatments include androgen suppression/deprivation methods that may be purely pharmaceutical (*e.g.,* hormonal manipulation with anti-androgens/LHRH agonists), surgical or mechanical ablative treatment (prostatectomy with orchiectomy, cryosurgery, or high-intensity focused ultrasound (HIFU)), and/or radiation treatment (*e.g.,* external beam radiation therapy (EBRT), proton beam irradiation, brachytherapy, radioimmunotherapy, targeted radiation) [20–32].

Over the past 40 years, a number of radiopharmaceuticals have been introduced to detect and treat prostate cancer by diagnostic imaging and targeted radiotherapy, respectively. The general principle for the development of these drugs is that they target malignant tissues and accumulate in sites of disease, often in areas of osteoblastic activity in bony metastases. More recently, radiolabeled peptides have been developed, which can target receptors on prostate tumors with high affinity and specificity. Receptors of interest for specific tumor targeting include prostate specific membrane antigen (PSMA), gastrin-releasing peptide receptor (GRPR), and $\alpha_v\beta_3$ integrin. Well-designed receptor-avid peptides have the ability to produce high imaging contrast and sensitivity for single photon emission computed tomography (SPECT) and positron emission tomography (PET), as well as a high therapeutic index for peptide receptor radionuclide therapy (PRRT), respectively.

8.2 Radiopharmaceutical Imaging and Therapy of Prostate Cancer

Radionuclides useful for peptide-based imaging and therapy of prostate cancer are listed in Table 8.1. Useful nuclear properties of radionuclides for imaging are gamma rays (γ) for scintigraphy and SPECT imaging and positrons (β^+) that produce annihilation radiation with an energy of 511 keV for PET imaging. Useful emissions for therapy include beta minus particles (β^-) and, less commonly, alpha particles (α^{++}).

Technetium-99m is called the "workhorse of nuclear medicine," as it is readily obtained onsite and on demand from a $^{99}Mo/^{99m}Tc$ generator in a hospital radiopharmacy, emits an optimal γ ray energy for nuclear imaging, and is used for approximately 85% of all imaging procedures. Much prostate cancer imaging is aimed at evaluating metastatic sites, particularly in bone. Planar γ scintigraphy using ^{99m}Tc-labeled methylenediphosphonate (^{99m}Tc-MDP) is currently the most commonly utilized imaging modality for prostate cancer staging, as well as for monitoring bony metastases prone to pathologic fracture. In addition, ^{99m}Tc-MDP is used to evaluate therapeutic response of disseminated disease. This radiocomplex targets the surface of remodeling bone in hydroxyapatite crystals during mineralization, becoming a useful tool for imaging metastatic bone lesions. In the case of cancer of prostatic origin, the skeletal metastases tend to exhibit osteoblastic expansion, as opposed to being characterized by osteoclastic bone destruction. Unfortunately, there are

Table 8.1 Properties of radionuclides used for peptide labeling

Radionuclide	Maximum particle emission (%)	Gamma emission (%)	Half-life	Application
[99m]Tc	– (IT)	140 keV (89.3%)	6.01 h	SPECT
[111]In	– (EC)	245 keV (94%) 171 keV (91%)	2.80 days	SPECT
[64]Cu	0.653 MeV β^+ (17.5%) 0.579 MeV β^- (38.5%) EC (43.5%)	1346 keV (0.5%) 511 keV (35%) ann. Rad.	12.7 h	PET
[68]Ga	1.899 MeV β^+ (89.1%) EC (10.5%)	1077 keV (2.99%) 511 keV (178%) ann. Rad.	67.7 min	PET
[86]Y	3.15 MeV β^+ (34.0%) EC (66.0%)	1921 keV (20.8%) 1854 keV (17.2%) 1153 keV (30.5%) 1077 keV (82.5%) 777 keV (22.5%) 703 keV (15.4%) 511 keV (67.2%) ann. Rad.	14.7 h	PET
[90]Y	2.28 MeV β^- (100%)	–	64.0 h	Targeted radiotherapy
[177]Lu	0.498 MeV β^- (78.6%) 0.385 MeV β^- (9.1%) 0.177 MeV β^- (12.2%)	208 keV (11.0%) 113 keV (6.4%)	6.73 days	Targeted radiotherapy

IT isomeric transition, *EC* electron capture, *ann. Rad.* annihilation radiation

multiple etiologies that can produce a false positive for bone metastasis, owing to the fact that accumulation may occur during normal bone remodeling or benign pathologies such as arthritis and general inflammation [33]. Nonetheless, bone scintigraphy with ^{99m}Tc-MDP is uniformly available and reimbursable through Medicare/Medicaid and is relatively inexpensive for whole-skeleton evaluation [34, 35]. Other modalities, such as PET using ^{18}F-fluoride (^{18}F-NaF$^-$), operate by the same mechanisms in bone and have the potential to be more sensitive. These drugs have shown superior imaging efficacy, accuracy, and spatial resolution due to the use of PET technology versus planar scintigraphy.

An early bone-targeting radiopharmaceutical developed and approved by the FDA to treat metastatic prostate cancer was [153]Sm-ethylenediaminetetramethylenephosphonate ([153]Sm-EDTMP, QuadraMet®). Like ^{99m}Tc-MDP, this agent accumulates in tumor sites of high osteoblastic activity, delivering internally emitting, cytotoxic β^- radiation. Recently, [223]RaCl$_2$ (Xofigo™), an α^{++}-emitting bone agent, was approved by the FDA. This drug is a calcium mimetic selective for osteoblastic repair. Both QuadraMet® and Xofigo™ have been shown to improve overall survival of prostate cancer patients and provide pain relief by controlling metastatic bone lesions. In some cases, tumor uptake of [153]Sm-EDTMP mimics that of ^{99m}Tc-MDP, enabling a routine bone scan to predict which patients are likely to benefit from QuadraMet® therapy. These drugs thus provide evidence that bone-seeking, therapeutic radiopharmaceuticals can also be very selective and useful.

8.3 Design of Radiolabeled Peptides for Imaging and Therapy

Radiolabeled peptides and small molecules continue to hold promise for early diagnosis and treatment of human disease [36]. The interest in radiolabeled peptides for molecular imaging and therapy stems from their rapid blood clearance, rapid urinary excretion, ease of penetration into tumor vasculature, relatively low immunogenicity, and their ability to be chemically tuned to target cell-surface receptors that tend to be expressed in high numbers on human cancer cells, relative to normal tissues [36]. The selective overexpression of peptide receptors in human cancers is often exploited in targeting strategies for radionuclide imaging and therapy (Fig. 8.1). The ability to use radiolabeled peptides to target cell surface receptors or biomarkers selectively has been a point of investigation for many years [37–41].

Design and development of radiolabeled, peptide-based, tumor-targeting agents for *in vivo* molecular imaging or therapy involves the preliminary consideration of many pieces of a complex puzzle. These pieces include a radiometal with appropriate emission characteristics, a metal complexing agent, a pharmacokinetic modifier, and a receptor-specific targeting vector. Important considerations for desirable radiopeptide pharmacology include blood clearance, receptor/biomarker binding kinetics, and excretion route. For molecular imaging using peptides, it is necessary for an agent to have its highest target uptake in a short period of time, as to achieve a useful diagnostic signal-to-noise ratio. Residence time of the tracer in blood should be minimal, yet long enough for the drug to reach the binding site of the biomarker. It is necessary for the agent to have rapid renal-urinary excretion, with minimal accumulation of radioactivity in the gastrointestinal tract, allowing for negligible irradiation of non-target tissues and diagnostically useful images of the lower abdomen.

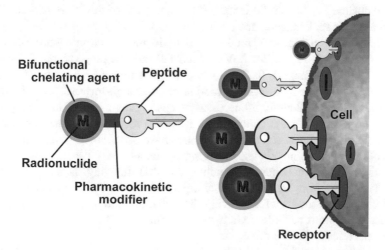

Fig. 8.1 Principle of peptide-based radiopharmaceutical targeting of cancer, based on Paul Ehrlich's 1897 "lock and key" theory of receptor binding

8.3.1   The Radionuclide

When choosing a radionuclide for use in a peptide-targeting construct, the main factors for consideration are the half-life, the emission profile, the method of production, and the means and ease by which the radioactive atom may be chemically attached. The half-life must be long enough to allow for synthesis, purification, transport, administration to the patient, and localization to target cells. At the same time the half-life must be short enough so as to prevent an excessive radiation dose to the patient. These characteristics make the half-life a matter of a few hours for optimal imaging and therapy. The emission profile must be one that is conducive to the task(s) intended, whether that be the need for γ photons for scintigraphy or SPECT, or for positron annihilation photons for PET imaging. For optimal imaging resolution, usable γ energies fall between 100–200 keV, while usable positron particle emissions are typically on the order of a few hundred keV. For therapy, beta minus or alpha emissions are generally necessary for efficacy. A combination of photon and cytotoxic emissions, allowing for concomitant imaging and therapy with the same radionuclide, is optimal for staging disease and evaluating response to targeted radiotherapy. Finally, the ideal radionuclide is readily available and not cost-prohibitive.

Indium-111-, ^{68}Ga-, ^{90}Y-, and ^{177}Lu-radiolabeled compounds have been of recent interest in peptide-based prostate cancer imaging and therapy, due to their availability and optimal nuclear characteristics. Indium-111, ^{68}Ga, and ^{177}Lu have half-lives, decay modes and emission profiles suitable for imaging and radiation dosimetry, and the β$^-$ emission of ^{177}Lu also renders it useful as a radiotherapeutic. Yttrium-90, on the other hand, cannot be used as a practical imaging agent, owing to the absence of γ ray emissions. However, it is suitable as a therapeutic radionuclide and can be used in a "matched pair" with another radionuclide with emissions useful for imaging and dosimetry. Yttrium-86 is the other component of this "matched pair," as it can be used for PET imaging and dosimetry.

Indium-111 is a cyclotron product that undergoes electron capture decay and emits two γ rays (171 and 245 keV), which fall into the range required for scintigraphic or SPECT imaging. Its 2.8-day half-life also makes ^{111}In optimal for molecular imaging. For example, since its production is straightforward, it can be readily obtained in time to allow for all necessary steps in the molecular imaging process, ranging from drug preparation to patient imaging procedures. Numerous monoclonal antibodies, peptides, and small molecules have been labeled with ^{111}In, making it one of the more commonly used imaging radionuclides.

Gallium-68, a promising radionuclide for PET imaging, is the product of a ^{68}Ge/^{68}Ga generator, which can be used onsite and on demand. It has a half-life of 68 min, which is attractive for reducing the radiation burden to an imaging patient. Because of its short half-life, ^{68}Ga is most useful to label rapidly targeting agents, such as peptides and small molecules.

The lanthanide and pseudolanthanide radiometals, such as ^{90}Y and ^{177}Lu, are produced directly or indirectly in a nuclear reactor and decay most often by β$^-$ emission, making them desirable for targeted radiotherapy. In addition, the radiolanthanides all possess very similar radiolabeling chemistries while offering a diverse spectrum of nuclear decay properties. Therefore, it is possible to match a desired set of nuclear properties, (e.g. half-life, maximum β^- energy, range) to a particular clinical application [42].

Yttrium-90 is produced most efficiently from a $^{90}Sr/^{90}Y$ generator, after production of ^{90}Sr in a nuclear reactor. Yttrium-90 has a half-life of 64 h, decays by β^- emission, and has a maximum particle range of approximately 11 mm in tissue, making it an attractive radionuclide for treatment of larger tumors [42, 43]. The utility of ^{90}Y lies in its pure β^- emission profile and this, in conjunction with its other characteristics, makes it a good therapeutic isotope choice. Lutetium-177 is produced directly in a nuclear reactor. It has a half-life of 6.73 days, decays by β^- emission, and emits two imageable γ photons (113 and 208 keV) that are useful for molecular imaging and tracking of administered doses in patients. In addition, the low-abundant γ photons allow the possibility of also determining the radiation dose to the patient. Lutetium-177 has a maximum particle range of approximately 2 mm in tissue, making it potentially useful for treating smaller tumors [42, 43]. Like ^{90}Y, it can be produced in large quantities and is an emerging radionuclide for therapeutic applications using peptide-based radiopharmaceuticals [44].

Targeted radiotherapy using ^{90}Y or ^{177}Lu is based on the use of tumor cell-seeking molecular vehicles that carry a therapeutic radionuclide to cancerous tissues, delivering a cytotoxic insult to tumor cells. This targeted approach has the potential for selective irradiation of diseased tissue while sparing adjacent normal tissues. An additional advantage of targeted radiotherapy is the ability to deliver a cytotoxic dose of radiation to collateral receptor-negative tumor cells, producing what is called a "cross-fire" effect.

8.3.2 The Metal Chelator

Without an appropriate metal chelator, the radionuclide may become disengaged from the peptide, resulting in an excessive imaging background signal, or worse, toxicity to non-target tissues. Other potential consequences of an inappropriate chelator choice or of a poor location for attachment include decreased binding affinity or selectivity. The choice of chelating agent depends mostly upon the properties of the radiometal it will contain and the ease of the conjugation chemistry. Kinetic and thermodynamic stability of the metal complex should also be considered. For example, demetallation from the chelator or transmetallation to serum proteins could result in nonspecific accumulation in collateral tissues. Such properties result in less than optimal target:non-target ratios, low-contrast imaging, and/or lower therapeutic indices with greater toxicity. Bifunctional chelating agents serve not only to contain and stabilize radionuclides, but also to tether them to other molecules via a chemically active functional group. These moieties must also be suitable for binding a given radiometal and be amenable to conjugation to a peptide, without causing undesirable changes in biodistribution, target binding and retention, and overall pharmacokinetics.

One of the most versatile metal complexing agents is DOTA (1,4,7,10-tetraaza-cyclododecane-1,4,7,10-tetraacetic acid), used to stabilize a variety of radiometals for imaging and therapy, such as ^{111}In for SPECT, ^{64}Cu and ^{68}Ga for PET, and ^{90}Y and ^{177}Lu for targeted radiotherapy. DOTA is one of several chelators with a macrocyclic structure. Cyclic moieties are generally superior to acyclic forms, as the large, ring-like moiety secures the radionuclide tightly at its center. DOTA can be conjugated to peptides through one of its carboxyl groups or at the carbon backbone of the ring.

8.3.3 The Linker

The final components of radiopharmaceutical peptides are the linking and spacers between the radiometal complex and the targeting vector; *i.e.*, peptide. These components also support the performance of the entire compound by influencing the final size, shape, solubility, stability, and molecular weight of the construct. Moreover, linking and spacing between the radiometal chelate and the peptide affect the pharmacodynamics, affinity, and mode of excretion of the bioconjugate. The pharmacological behavior of tracer peptide-based agents can also be affected by chemical modification of the metal-chelate complex and/or biomarker binding region or insertion of a linking moiety between the two as a pharmacokinetic modifier [45]. Insertion of amino acid or aliphatic linkers between the binding region of peptides and the metal-chelate complex usually does not reduce receptor binding affinity significantly [46, 47]. Shorter, more-polarizable linkers tend to render a more hydrophilic radioligand with excretion primarily *via* the renal-urinary pathway. Long-chained aliphatic linkers, on the other hand, tend to produce more hydrophobic radiopeptides with unfavorably slow clearance *via* the hepatobiliary pathway. Desirable peptide-based targeting radiopharmaceuticals will bind target receptors with high affinity, remain stable under normal physiological conditions, penetrate tumors and tumor vasculature, be quickly cleared from the blood and non-target tissues, and have minimal immunogenicity [36–38, 41, 48]. Because tumor pathology is targeted by radiolabeled peptides, there is potential for using these compounds to detect not only primary tumor locations, but also areas of metastasis as well. They may be used to evaluate and stage a patient with one test, providing information helpful in the process of therapy selection and approaches.

The characteristics of radiolabeled peptides provide strong motive for the continued development and refinement of this category of diagnostic and therapeutic agents. There is much opportunity for investigating the numerous, unexplored combinations of their structural components that favorably alter their pharmacokinetic profiles. While there are novel diagnostic and therapeutic peptides being developed, it is important to build upon the progress and investment made thus far.

8.4 Targeting Prostate Specific Membrane Antigen

Some of the most promising imaging agents currently in evaluation target PSMA. In prostate cancer, it is quite overexpressed (100- to 1000-fold more than in normal tissues such as the brain, kidney and small intestine) [49]. Moreover, PSMA tends to increase with the degree of aggressiveness and metastatic potential of prostate cancer [49, 50].

Prostascint™ is an FDA-approved murine monoclonal antibody labeled with ^{111}In for scintigraphy, which was used extensively for many years to target prostate-specific membrane antigen (PSMA). Unfortunately Prostascint™ reacts with an intracellular epitope of PSMA that is only exposed in necrotic cells. Thus, this drug was reported to be less sensitive and specific than agents developed since, such as receptor-targeting peptides and small molecules.

Most imaging agents targeted to PSMA are small molecule enzyme inhibitors aimed at the catalytic domain of the receptor, reviewed by Pillai et al. [51]. These compounds have been labeled with ^{99m}Tc for scintigraphy and SPECT imaging, ^{18}F and ^{68}Ga for PET imaging, as well as ^{177}Lu for targeted radiotherapy. The PSMA-targeting probe DUPA (2-[3-(1, 3-dicarboxy propyl)-ureido] pentanedioic acid) has often been investigated for prostate cancer targeting. Gallium-68-PSMA-11, a DUPA derivative developed by Afshar-Oromieh et al. [52, 53], is a PET imaging agent currently in clinical trials for detection of PSMA on resected prostate tissue performed in high-risk cases. Studies to evaluate, characterize, and compare it to bone scintigraphy with ^{99m}Tc suggest that ^{68}Ga-PSMA-11 for PET imaging, combined with CT or MRI, is at least comparable if not superior to scintigraphy. Scintigraphy will only detect bone lesions, whereas PSMA-11 can detect primary tumors and distant soft tissue metastases, in addition to bone lesions [52, 54–56]. This advantage was apparent particularly with regard to sensitivity for finding recurrent prostate cancer in a low serum PSA scenario.

8.5 Targeting Gastrin-Releasing Peptide Receptor

It is well-established that gastrin-releasing peptide (GRP) is a potent epithelial mitogen and that its receptor, gastrin-releasing peptide receptor (GRPR), also known as bombesin (BBN) receptor subtype 2 or BBN2, has been identified in tissue biopsy samples and immortalized cell lines of many human cancers [57–59]. For example, GRPRs are present in high concentrations on prostate, breast, pancreatic, and small cell lung carcinomas [60].

The roles of the GRPR *in vivo* include the regulation of several physiologic processes and effecting a wide range of body systems in an autocrine fashion [61]. Activation of the GRPR stimulates phospholipase C (PLC) and results in activation of tyrosine and mitogen-activated protein (MAP) kinase cascades. In castrate-resistant prostate cancer, activation of the GRPR stimulates growth by upregulating cyclooxygenase-2 (COX-2) and releasing prostaglandin E2 (PGE2) *via* activation of the P13K/AKT pathway and MAP kinase p38, which prevents the degradation of COX-2 [62, 63]. Even in the absence of androgens, androgen receptors (AR) can be recruited and upregulated by tyrosine phosphorylation *via* the GRPR [62]. The vasculature of many different human tumors reflects high expression of the GRPR concurrently with vascular endothelial growth factor (VEGF), which indicates a role in angiogenesis to support growth and metastasis [64].

Mammalian GRPRs are G protein-coupled, 7-transmembrane receptors having the capacity to be endocytosed upon binding by an effective agonist ligand. There are four known BBN receptor subtypes [65]: BB1 (neuromedin B receptors, NMBR), BB2 (GRPR), BB3 (BRS-3), and BB4 (BRS-4). Early on, the ability of GRPR-targeting ligands containing a BBN receptor-binding motif to be rapidly internalized, coupled with a high incidence of receptor expression on various tumors, led to the design and development of new diagnostic and therapeutic agents targeting the GRPR as a clinical biomarker for early detection, staging, and potential treatment of cancer.

The literature reveals the extent to which overexpression of the GRPR occurs in prostate cancer. Markwalder and Reubi performed studies that revealed the following: presence of GRPR expression in primary prostatic invasive carcinoma in all cancerous tissues assayed, with 83% of those determined to be high or very high density (1000 dpm/mg tissue); 25 out of 26 patients with prostatic intraepithelial neoplasia (PIN) demonstrated high to very high densities of GRP receptors; and, more than half of the castrate resistant prostate cancer bone metastases were GRPR-positive [57]. These findings are consistent with the findings of a study ten years later by Ananias et al. [66]. It is also important to note that these analyses in non-neoplastic prostatic tissues and in BPH cases showed GRPR to be either completely absent or of negligible incidence [57]. Near the same time, Schally and coworkers confirmed the preferential expression of GRPRs in prostatic carcinoma, with 91% of the samples expressing mRNA for the GRPR [59].

The model of success in the development of receptor-avid peptides has been the targeting of somatostatin receptors for the development of both diagnostic and therapeutic radiopharmaceuticals [41]. For example, the successes of OctreoScan® (^{111}In-DTPA-Octreotide) have paved the way for radiolabeled peptide exploration of other receptor systems. These peptide/receptor systems include bombesin receptor subtype 2 (BBN2), alpha-melanocyte stimulating hormone (α-MSH), arginine-glycine-aspartic acid (RGD), vasoactive intestinal peptide (VIP), cholecystokinin, and neurotensin (NT) [37, 38, 67].

Bombesin is a 14-amino acid amphibian peptide analogue (originally derived from the oriental fire-bellied toad, *Bombina orientalis*) [68] of the 27-amino acid mammalian regulatory peptide GRP. BBN and GRP share a homologous 7-amino acid amidated *C*-terminus, Trp-Ala-Val-Gly-His-Leu-Met-NH$_2$, (BBN(7–14)NH$_2$, Fig. 8.2), which is essential for high-affinity receptor binding to the GRPR [46, 47]. BBN-based molecules that target the GRPR in an agonistic fashion mimic the endogenous gastrin-releasing peptide and are among the most utilized ligands for targeting human cancers. Upon binding to the GRPR, these peptides are rapidly internalized in tumor cells.

Numerous researchers have employed radiolabeled, bombesin-based ligands to target prostate cancer for imaging, because of their high binding affinity for the GRPR. These studies account for an abundance of literature pertaining to the subject. This literature has been reviewed by Baratto et al. [69]. Radiolabeled agonists, based on the *C*-terminal 7–14 amino acids, were the first peptides developed for imaging of the GRPR [70–74]. The first agents were labeled with ^{99m}Tc [75, 76], but while these peptides showed tumor localization with high specificity, clinical studies were only performed in small cohorts of patients. The performance of these peptides has not yet been studied in larger cohorts, in order to make a better determination of their potential utility.

Fluorine-18- and ^{64}Cu-labeled agonists were next developed and studied in cell and animal models of prostate cancer. Dijkgraaf et al. [77] labeled an aluminum-chelated NOTA (1,4,7-triazacyclononane-1,4,7-triacetic acid) with ^{18}F and found that its uptake in PC-3 tumors xenografted to mice was specific for GRPR. PC-3 is the most commonly used mouse prostate model, which is based on a skeletal metastasis characteristic of advanced disease. More recently, Carlucci et al. [78] developed two BBN analogues labeled with ^{18}F that showed high specificity and stability in prostate tumor-bearing mice.

Copper-64-labeled agonists were being developed concurrently. Lane et al. labeled a NOTA-based BBN(7–14)NH$_2$ conjugate with ^{64}Cu and evaluated a series of pharmacokinetic-modifying linkers between the chelate and the *N*-terminus of the peptide [79]. They found that an eight-carbon linker (Fig. 8.2, where M = ^{64}Cu) produced the highest tumor uptake and fastest renal clearance in tumor-bearing mice. They also obtained micro-PET/CT images (Fig. 8.3) that clearly delineated PC-3 tumors. However, the images also showed high liver uptake characteristic of ^{64}Cu. Copper transchelates to superoxide dismutase in the liver [80], causing excessive and persistent accumulation. To alleviate this problem, bombesin-based imaging agents might be labeled with ^{68}Ga in the future. Other groups have studied

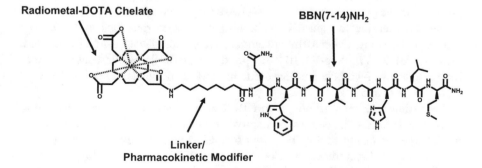

Fig. 8.2 Structure of radiometal-labeled BBN(7–14)NH$_2$

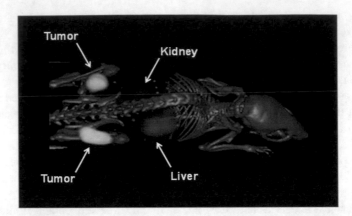

Fig. 8.3 Micro-PET image of a PC-3 prostate tumor-bearing mouse, using ^{64}Cu-labeled BBN(7–14)NH$_2$

^{64}Cu-labeled agonists, also finding high specificity and stability *in vivo* [81, 82]. As of yet, however, ^{18}F- and ^{64}Cu-labeled agents have not been studied in humans.

GRPR antagonists have received much attention in recent years. Antagonists bind to the receptor without internalizing or activating it, bind to a higher number of sites per receptor, successfully target the receptor, and are retained in GRPR-positive tumors. The antagonist RM1, utilizing a statine residue, was labeled with ^{111}In and compared to the corresponding agonist, AMBA, by Mansi et al. [83]. In PC-3-bearing mice, they found that the antagonist displayed higher tumor uptake and higher tumor-to-normal tissue ratios than the agonist. In addition, the ^{68}Ga-labeled antagonist displayed similar pharmacokinetics to the ^{111}In peptide, giving rise to the possibility of PET/CT imaging of GRPR in the clinic. The antagonist RM2, 4-amino-1-carboxymethyl-piperidine-D-Phe-Gln-Trp-Ala-Val-Gly-His-Sta-Leu-NH$_2$, was developed by the same group in 2011 [84]. Like RM1, when conjugated to an appropriate radiometal chelation moiety, it has been shown to have improved uptake and retention in tumors as compared to agonistic-type GRPR-targeting ligands. Indium-111- and ^{68}Ga-RM2 also showed highly specific targeting of PC-3 tumors in mice; however, abdominal uptake was observed, primarily in the pancreas, which also expresses high levels of GRPR. Radioactivity in the abdomen cleared quickly, while tumor retention remained high over the course of the study. Other BBN antagonists have been developed and evaluated, including MJ9 [85], NeoBOMB1 [86], and MATBBN [87–90]. These peptides have also shown favorable pharmacokinetics and tumor uptake when labeled with ^{64}Cu, ^{68}Ga, ^{18}F, and ^{111}In.

Promising preclinical results, with GRPR antagonists in particular, have led to clinical trials for staging and restaging prostate cancer patients. Gallium-68-labeled RM2 was evaluated in 14 men [91] with scheduled radical prostatectomies (11) or with biochemical recurrence (3). Comparison of PET/CT with histology as a gold standard showed that ^{68}Ga-RM2 had high sensitivity (89%), specificity (81%), and accuracy (83%) for detecting primary disease. The antagonist AR06 was labeled with ^{64}Cu and evaluated in 4 patients newly diagnosed with prostate cancer [92], affording clear, high-contrast detection of tumors. A recent restaging study compared ^{68}Ga-RM2 to ^{18}F-fluoroethylcholine (^{18}F-ECH) in 16 men with biochemical recurrence and negative or inconclusive ECH scans [93]. PET/CT of ten patients showed abnormal RM2 uptake. In two patients with inconclusive or negative ECH scans, RM2 detected additional pelvic lymph nodes and bone metastases. However, PSA levels in the patients receiving ECH scans were lower than those at the subsequent time of a RM2 imaging, necessitating larger prospective trials to determine if PSA is a factor in restaging with these agents. An interesting pilot study [94] compared ^{68}Ga-labeled RM2 to ^{68}Ga-PSMA-11 in seven patients with castration-resistant prostate cancer and ambiguous results by conventional imaging (such as bone scintigraphy, CT, and/or MRI). Despite the fact that two different receptors were targeted, GRPR and PSMA respectively, similar uptake patterns were obtained for the two imaging agents. The same group [95] subsequently reported a prospective study in which 32 similar patients underwent PET/MRI imaging with ^{68}Ga-labeled RM2. In this trial, ^{68}Ga-RM2 PET detected more cases of recurrent disease (23) than MRI (11).

8.6   Targeting $\alpha_v\beta_3$ Integrin

More recent targets of interest include members of the integrin receptor family suspect in the development of metastasis due to roles in cell attachment, cell-to-extracellular matrix interactions and angiogenesis, and the fact that they are expressed in many tumors, including malignant melanoma, osteosarcoma, glioblastoma, and breast as well as prostate carcinomas [96–98]. Integrins are a class of heterodimeric, glycoprotein cell adhesion molecules (CAMs) that control the connection of cellular cytoskeletons to attach to other cells or the extracellular matrix. In this way, they regulate cell adhesion, migration, proliferation and differentiation and, thus, angiogenesis and metastasis [98–104]. Mammals possess 18 different α subunits and eight different β subunits, which enables assembly of at least 24 functional combinations [103].

Integrins are cell-surface transmembrane glycoproteins. The $\alpha_v\beta_3$ and $\alpha_v\beta_5$ integrin subtypes are expressed on the endothelial cells of tumor neovasculature during angiogenesis and form the basis of investigations for molecular imaging of angiogenesis and tumor formation in $vivo$. The $\alpha_v\beta_3$ integrin is known to be expressed in very high numbers in many tumor types including lung carcinomas, osteosarcomas, breast cancer, and glioblastomas [105]. It is also over-expressed in some prostate tumors at different stages [106, 107]. Radiolabeled peptides containing the amino acid sequence Arg-Gly-Asp (RGD) are non-regulatory peptides that have been used extensively to target $\alpha_v\beta_3$ receptors upregulated on tumor cells and neovasculature. These peptides therefore provide a molecular vehicle for early detection of rapidly growing tumors and metastatic disease.

RGD peptides bind between the alpha and beta subunits of the integrin, as evidenced by crystallography [108]. Cyclization of the RGD sequence prolongs its biological half-life, increases selectivity and binding affinity, and provides a site for conjugation at the site of a lysine residue [109, 110]. RGD conjugated to chemotherapeutics is the subject of many studies with the goal of providing site-directed killing of cancer cells to reduce exposure and toxicity to normal, non-target tissues [111–114]. It has been documented that multimeric use of the cyclic RGD motif increases binding affinity, tumor uptake, and retention [109, 115]. Monomeric and multimeric RGD peptides have been used as prostate cancer molecular imaging agents and have shown the ability to target $\alpha_v\beta_3$ integrin specifically *in vivo* [98, 109, 116, 117]. Several RGD-based peptides have been labeled with radionuclides such as ^{99m}Tc, ^{111}In, ^{68}Ga, ^{18}F, and ^{64}Cu and evaluated for imaging tumors that express $\alpha_v\beta_3$ [105, 109, 112, 115, 117–123]. Tumor targeting and uptake of these peptide conjugates are mediated by both $\alpha_v\beta_3$ expression levels and activation levels [124].

Israel et al. [125] compared the prostate cancer imaging efficacy of ^{68}Ga-labeled RGD to those of ^{18}F-fluorodeoxyglucose (FDG) and ^{18}F-fluoroethylcholine in PC-3 and DU-145 prostate tumor-bearing mice. While uptake of ^{68}Ga-RGD was lower than those of the other two drugs, it afforded higher tumor-to-background ratios in muscle and bone. The clear detection of tumors shown in micro-PET images of the mice identified the peptide as a promising candidate for prostate cancer imaging.

Efforts to enhance tumor uptake of RGD conjugates have focused primarily on increasing avidity by synthesizing dimer and tetramer constructs of the peptide. Hu and coworkers [126] prepared an ^{18}F-labled RGD dimer and evaluated it in prostate tumor-bearing mice. Blood clearance was rapid, and excretion of peptide occurred predominantly through the renal pathway. These characteristics led to high tumor imaging contrast and identified this peptide as a potential PET imaging agent.

Cheng et al. evaluated another ^{18}F-labeled RGD dimer and compared its imaging capabilities to FDG, ^{18}F-fluoro-3′-deoxy-3′-L-flourothymidine (FLT), and ^{18}F-fluoromisonidazole (FMISO) in nude mice bearing prostate cancer xenografts [127]. FDG had the highest tumor uptake and tumor-to-muscle ratio, albeit with high uptake in the brain, heart, and intestinal tract. FLT and FMISO showed modest accumulation in the tumor, but lower uptakes in physiologically collateral tissues. While the tumor uptake of the RGD peptide was the lowest, it still demonstrated high tumor specificity.

These preclinical studies identified RGD peptides as potential candidates for imaging prostate cancer patients. A proof-of-concept study by Schwarzenböck and coworkers [128] evaluated ^{18}F-Galacto-RGD in 12 prostate cancer patients. In this study, PET imaging using this peptide produced tumor-to-blood/tumor-to-muscle ratios of 1.4/2.8 for bone metastases, 1.5/3.2 for malignant lymph nodes, and 2.0/3.9 for primary tumors. Of the 74 bone lesions identified by conventional scintigraphy, 58 were also detected by ^{18}F-Galacto-RDG PET imaging. Because of its rapid clearance from physiologic tissues, the peptide was able to detect bone metastases with low uptake of the imaging agent, affording high sensitivity. Excretion of the peptide was predominantly through the renal pathway, but there was also some hepatobiliary clearance, which could obscure lymph nodes in the abdomen. Of the five patients with lymph node metastases, uptake could be confirmed in two, but both positive and negative nodes had accumulation of ^{18}F-Galacto-RDG, indicating a lack of specificity. Primary tumors in six of seven patients were detected, although imaging was hindered by the high degree of urinary excretion. While ^{18}F-Galacto-RDG PET imaging was inferior to bone scintigraphy, it is possible that patients could be screened as candidates for $\alpha_v\beta_3$-targeted therapy.

8.7   Recent Advancements

8.7.1   Peptide–Receptor Radionuclide Therapy of Prostate Cancer

One of the most important emerging applications of peptide-based tumor targeting is translation of molecular imaging agents to peptide-receptor radionuclide therapy (PRRT). In recent years, several cancer imaging antibodies and peptides containing γ-emitting radionuclides, such as ^{111}In and ^{99m}Tc, have been labeled with other radiometals that emit cytotoxic charged particles, typically β$^-$ particles. In the United States, the leading agent used for clinical PRRT is ^{177}Lu-DOTA-Tyr^3-octreotate (Lutathera®), which was approved by the FDA in January 2018 for somatostatin receptor-targeted radiotherapy of patients with neuroendocrine tumors. In this case, the ^{111}In-DTPA chelate on OctreoScan® was replaced with DOTA chelating ^{177}Lu.

For prostate cancer PRRT, bombesin peptides have been investigated in the laboratory for targeting BB2. Lantry and co-workers first described the synthesis, characterization, in vitro, and in vivo studies of ^{177}Lu-AMBA in 2004 [129]. AMBA (DOTA-CH$_2$CO-Gly-4-aminobenzoyl-Gln-Trp-Ala-Val-Gly-His-Leu-Met-NH$_2$) is an agonist peptide agent based upon the BBN(7–14)NH$_2$ targeting motif and was described to have selective binding to the BB1 (NMBR) and BB2 (GRPR) receptors [129, 130]. Using World Health Organization criteria, the authors concluded that the usage of two dose administrations (55.5 MBq total) of ^{177}Lu-AMBA improved the median survival of prostate tumor-bearing mice by 36%, and time to progression/progression-free survival increased by 65%, as compared to administration of only a single dose (27.75 MBq). For example, overall survival at the 30, 60, 90, and 120 day time points using a single-dose administration was 100%, 97%, 47%, and 38%, respectively. For the two-dose strategy, overall survival was 100%, 100%, 78%, and 47% at the same time-points [129]. The group at Bracco S.p.A. conducted Phase I clinical trials in human patients. Bodei and co-workers reported the first Phase I human clinical trial in 2007 [131]. In this study, seven patients presenting with metastatic hormone refractory prostate cancer (HRPC) were studied and evaluated for safety, biodistribution, and dosimetry. Initially, all seven patients received a dose administration ranging from 1.14 to 1.94 GBq. Positive uptake of tracer was identified in five of the seven patients. Subsequently, three patients presenting with very high tumor uptake of tracer received a second dose of ^{177}Lu-AMBA in the range of 1.47–2.92 GBq to complete the study. The authors concluded that administration of ^{177}Lu-AMBA was safe. Side effects included abdominal cramps, diarrhea, and nausea. All of the patients presented with uptake in normal pancreatic tissue.

8.7.2   Dual-Receptor Targeting of Prostate Cancer

In recent years, investigators have been developing bivalent prostate cancer-targeting peptides that have the potential to target two receptors. Of particular interest are peptides with both gastrin-releasing peptide receptor and $\alpha_v\beta_3$ binding motifs; namely, chimeric BBN-RGD targeting vectors. It should be noted that it is highly unlikely that these peptides are capable of binding events that crosslink the two receptors; rather, they likely distribute between the two within tumors. Thus, the current thinking regarding the use of these peptides is twofold. First, it is possible that, if the peptide has targeting capabilities for both receptors, then more tumors can be detected, given that they express either GRPR or $\alpha_v\beta_3$, or both. Second, $\alpha_v\beta_3$ is a marker for angiogenesis in early tumors, and GRPR is associated with metastatic disease. Therefore, it may be possible to stage prostate cancer patients by molecular imaging using these bivalent peptides.

Chen and coworkers [132] were the first to report preparation of a bivalent BBN-RGD peptide, labeled with ^{18}F. This peptide bound specifically to both receptors *in vitro* and showed higher tumor uptake in PC-3 tumor-bearing mice than either corresponding monovalent targeting vector. However, this compound suffered from high liver uptake, which compromised the contrast of micro-PET images of xenograft-bearing mice. This bivalent peptide has recently been labeled with ^{68}Ga and translated to the clinic [133]. In 13 patients, ^{68}Ga-BBN-RGD detected three primary tumors, 14 metastatic lymph nodes, and 20 skeletal metastases.

Jackson et al. prepared a ^{64}BBN-RGD construct and evaluated it in the same cell and mouse models [105]. Tumor uptake in mice was high, as was retention of radioactivity in the xenograft. In contrast to the ^{18}F-labeled peptide, renal excretion was rapid, and uptakes in other organs were low with the exception of the BB2-rich pancreas. Around this time, a ^{177}Lu-labeled BBN-RGD peptide with potential for PRRT was synthesized by Cheng and colleagues [127]. This compound also showed high PC-3 tumor accumulation and retention, as well as high tumor-to-blood and tumor-to-muscle ratios, demonstrating its potential for effective therapy.

However, a potential problem concerning the bispecific properties of these conjugates exists in the fact that they are constructed around a GRPR agonist. As the agonist moiety is rapidly internalized upon binding to BB2, the peptide may not be available for binding $\alpha_v\beta_3$, which is expressed on the vascular endothelium. Therefore, Durkan et al. [134] and Stott Reynolds et al. [135] developed a chimeric peptide based on RGD and RM2, the non-internalizing GRPR antagonist. The structure of this bivalent peptide is shown in Fig. 8.4. In the first study, the conjugate was labeled with ^{64}Cu for PET imaging. Micro-PET images of PC-3 tumor-bearing

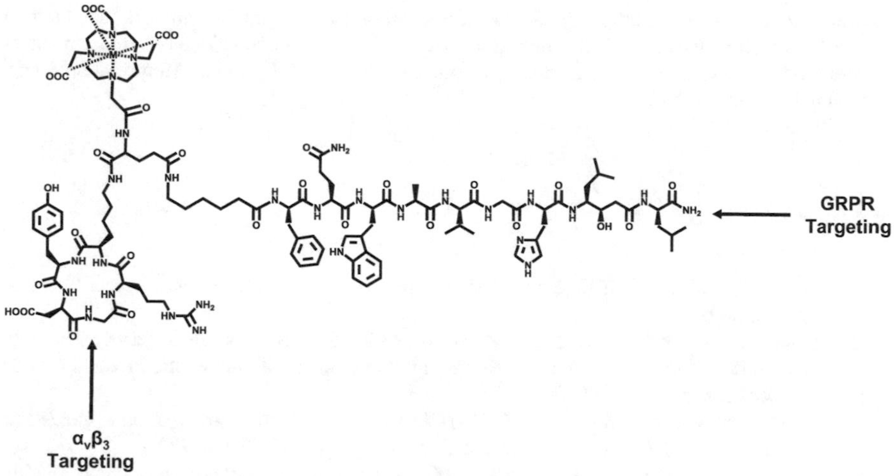

Fig. 8.4 Structure of a radiometal-labeled bivalent peptide targeting gastrin-releasing peptide receptor and $\alpha_v\beta_3$ integrin

mice produced high contrast images with minimal uptake in normal organs, with the exception of a small amount of radioactivity in the abdomen. In the latter report, ^{111}In for SPECT imaging was evaluated, and again tumors were detected with high contrast and very little accumulation in normal organs. The biodistribution of the ^{177}Lu-labeled peptide showed high tumor uptake, rapid urinary excretion, and low uptakes in normal organs. Furthermore, radioactivity in the pancreas washed out rapidly. These properties make this radiopeptide an attractive candidate for PRRT.

8.8 Closing Remarks

Radiopharmaceutical imaging and therapy of prostate cancer has progressed greatly in the last 40 years. The imaging agents initially used, such as ^{99m}Tc-MDP, were tissue-targeting drugs aimed at the osteoblastic activity of skeletal metastases. More recently, bone-seeking radiopharmaceuticals have been developed for PET imaging (^{18}F-NaF[-]) and targeted radiotherapy ([153]Sm-EDTMP, [223]RaCl$_2$), but these drugs still operate on the same mechanisms of uptake in osteoblastic bone tissue. Now peptide-based radiopharmaceuticals have ushered in a new era of versatile molecular targeting of prostate cancer, with the promise that tumors and metastatic lesions may be detected with increasing sensitivity and specificity. The advancements made in peptide-receptor molecular imaging and therapy, for both early and advanced disease, create new possibilities of more effective, personalized medicine for prostate cancer patients.

Acknowledgments The authors thank Donald Connor for the graphic artwork depicted in Fig. 8.1, as well as Jade Jones for editorial assistance. We also acknowledge the Department of Veterans Affairs, for the use of facilities and resources at the Harry S. Truman Memorial Veterans' Hospital in Columbia, MO.

References

1. Howlader N et al (2016) SEER cancer statistics review, 1975–2014. National Cancer Institute, Bethesda, MD
2. Jayasekera J, Onukwugha E, Bikov K, Mullins CD, Seal B, Hussain A (2014) The economic burden of skeletal-related events among elderly men with metastatic prostate cancer. PharmacoEconomics 32:173–191
3. Yong C, Onukwugha E, Mullins CD (2014) Clinical and economic burden of bone metastasis and skeletal-related events in prostate cancer. Curr Opin Oncol 26:274–283
4. Attar RM, Takimoto CH, Gottardis MM (2009) Castration-resistant prostate cancer: locking up the molecular escape routes. Clin Cancer Res 15:3251–3255
5. Harris WP, Mostaghel EA, Nelson PS, Montgomery B (2009) Androgen deprivation therapy: progress in understanding mechanisms of resistance and optimizing androgen depletion. Nat Clin Pract Urol 6:76–85
6. Croswell JM, Kramer BS, Crawford ED (2011) Screening for prostate cancer with PSA testing: current status and future directions. Oncology (Williston Park) 25:452–460 463
7. D'Amico AV (2012) Prostate-cancer mortality after PSA screening. N Engl J Med 366:2229 author reply 2230–2221
8. Duffy MJ (2011) Prostate-specific antigen: does the current evidence support its use in prostate cancer screening? Ann Clin Biochem 48:310–316
9. Hayes JH, Barry MJ (2014) Screening for prostate cancer with the prostate-specific antigen test: a review of current evidence. JAMA 311:1143–1149
10. Henson DE, Siddiqui H, Schwartz AM (2010) Re: overdiagnosis in cancer. J Natl Cancer Inst 102:1809–1810 author reply 1810–1801
11. Howrey BT, Kuo YF, Lin YL, Goodwin JS (2013) The impact of PSA screening on prostate cancer mortality and overdiagnosis of prostate cancer in the United States. J Gerontol A Biol Sci Med Sci 68:56–61
12. Stampfer MJ, Jahn JL, Gann PH (2014) Further evidence that prostate-specific antigen screening reduces prostate cancer mortality. J Natl Cancer Inst 106:dju026. https://doi.org/10.1093/jnci/dju026
13. Zappa M et al (2014) A different method of evaluation of the ERSPC trial confirms that prostate-specific antigen testing has a significant impact on prostate cancer mortality. Eur Urol 66:401–403
14. Albertsen PC, Hanley JA, Barrows GH, Penson DF, Kowalczyk PD, Sanders MM, Fine J (2005) Prostate cancer and the will Rogers phenomenon. J Natl Cancer Inst 97:1248–1253
15. Chan TY, Partin AW, Walsh PC, Epstein JI (2000) Prognostic significance of Gleason score 3+4 versus Gleason score 4+3 tumor at radical prostatectomy. Urology 56:823–827
16. Thompson IM, Canby-Hagino E, Lucia MS (2005) Stage migration and grade inflation in prostate cancer: will Rogers meets garrison Keillor. J Natl Cancer Inst 97:1236–1237
17. Albertsen PC, Moore DF, Shih W, Lin Y, Li H, Lu-Yao GL (2011) Impact of comorbidity on survival among men with localized prostate cancer. J Clin Oncol 29:1335–1341
18. Lu-Yao GL et al (2009) Outcomes of localized prostate cancer following conservative management. JAMA 302:1202–1209
19. Prostate (2010) In: Edge S, Byrd DR, Compton CC, Fritz AG, Greene F, Trotti A (eds) AJCC cancer staging handbook, 7th edn. Springer, New York, NY, pp 457–468

20. Aus G, Pileblad E, Hugosson J (2002) Cryosurgical ablation of the prostate: 5-year follow-up of a prospective study. Eur Urol 42:133–138
21. Chan TY, Tan PW, Tang JI (2016) Proton therapy for early stage prostate cancer: is there a case? OncoTargets Ther 9:5577–5586
22. Chaussy CG, Thüroff S (2017) High-intensity focused ultrasound for the treatment of prostate cancer: a review. J Endourol 31:S30–S37
23. Forman JD, Zinreich E, Lee DJ, Wharam MD, Baumgardner RA, Order SE (1985) Improving the therapeutic ratio of external beam irradiation for carcinoma of the prostate. Int J Radiat Oncol Biol Phys 11:2073–2080
24. Group Prostate Cancer Trialists' Collaborative (2000) Maximum androgen blockade in advanced prostate cancer: an overview of the randomised trials. Lancet 355:1491–1498
25. Parmar H, Edwards L, Phillips RH, Allen L, Lightman SL (1987) Orchiectomy versus long-acting D-Trp-6-LHRH in advanced prostatic cancer. Br J Urol 59:248–254
26. Peeling WB (1989) Phase III studies to compare goserelin (Zoladex) with orchiectomy and with diethylstilbestrol in treatment of prostatic carcinoma. Urology 33:45–52
27. Ploysongsang SS, Aron BS, Shehata WM (1992) Radiation therapy in prostate cancer: whole pelvis with prostate boost or small field to prostate? Urology 40:18–26
28. Sartor O, Hoskin P, Bruland OS (2013) Targeted radio-nuclide therapy of skeletal metastases. Cancer Treat Rev 39:18–26
29. Tyson MD, Penson DF, Resnick MJ (2017) The comparative oncologic effectiveness of available management strategies for clinically localized prostate cancer. Urol Oncol 35:51–58
30. Vale CL et al (2016) Addition of docetaxel or bisphosphonates to standard of care in men with localised or metastatic, hormone-sensitive prostate cancer: a systematic review and meta-analyses of aggregate data. Lancet Oncol 17:243–256
31. Welch HG, Albertsen PC (2009) Prostate cancer diagnosis and treatment after the introduction of prostate-specific antigen screening: 1986–2005. J Natl Cancer Inst 101:1325–1329
32. Zlotta AR et al (2013) Prevalence of prostate cancer on autopsy: cross-sectional study on unscreened Caucasian and Asian men. J Natl Cancer Inst 105:1050–1058
33. O'Sullivan GJ, Carty FL, Cronin CG (2015) Imaging of bone metastasis: an update. World J Radiol 7:202–211
34. Iagaru AH, Mittra E, Colletti PM, Jadvar H (2016) Bone-targeted imaging and radionuclide therapy in prostate cancer. J Nucl Med 57:19S–24S
35. Raval A, Dan TD, Williams NL, Pridjian A, Den RB (2016) Radioisotopes in management of metastatic prostate cancer. Indian J Urol 32:277–281
36. Reubi JC, Maecke HR (2008) Peptide-based probes for cancer imaging. J Nucl Med 49:1735–1738
37. Behr TM, Gotthardt M, Barth A, Béhé M (2001) Imaging tumors with peptide-based radioligands. Q J Nucl Med 45:189–200
38. Blok D, Feitsma RI, Vermeij P, Pauwels EJ (1999a) Peptide radiopharmaceuticals in nuclear medicine. Eur J Nucl Med 26:1511–1519
39. Blum J, Handmaker H, Rinne NA (2002) Technetium labeled small peptide radiopharmaceuticals in the identification of lung cancer. Curr Pharm Des 8:1827–1836
40. Blum JE, Handmaker H (2002) Small peptide radiopharmaceuticals in the imaging of acute thrombus. Curr Pharm Des 8:1815–1826
41. Kwekkeboom D, Krenning EP, de Jong M (2000) Peptide receptor imaging and therapy. J Nucl Med 41:1704–1713
42. Cutler CS, Smith CJ, Ehrhardt GJ, Tyler TT, Jurisson SS, Deutsch E (2000) Current and potential therapeutic uses of lanthanide radioisotopes. Cancer Biother Radiopharm 15:531–545
43. Smith CJ et al (2003a) Radiochemical investigations of ^{177}Lu-DOTA-8-Aoc-BBN[7-14]NH2: an in vitro/in vivo assessment of the targeting ability of this new radiopharmaceutical for PC-3 human prostate cancer cells. Nucl Med Biol 30:101–109

44. Li WP, Smith CJ, Cutler CS, Hoffman TJ, Ketring AR, Jurisson SS (2003) Aminocarboxylate complexes and octreotide complexes with no carrier added ^{177}Lu, ^{166}Ho and ^{149}Pm. Nucl Med Biol 30:241–251
45. Liu S, Edwards DS (1999) 99mTc-labeled small peptides as diagnostic radiopharmaceuticals. Chem Rev 99:2235–2268
46. Smith CJ, Volkert WA, Hoffman TJ (2003b) Gastrin releasing peptide (GRP) receptor targeted radiopharmaceuticals: a concise update. Nucl Med Biol 30:861–868
47. Smith CJ, Volkert WA, Hoffman TJ (2005) Radiolabeled peptide conjugates for targeting of the bombesin receptor superfamily subtypes. Nucl Med Biol 32:733–740
48. Reubi JC (2003) Peptide receptors as molecular targets for cancer diagnosis and therapy. Endocr Rev 24:389–427
49. Silver DA, Pellicer I, Fair WR, Heston WD, Cordon-Cardo C (1997) Prostate-specific membrane antigen expression in normal and malignant human tissues. Clin Cancer Res 3:81–85
50. Bostwick DG, Pacelli A, Blute M, Roche P, Murphy GP (1998) Prostate specific membrane antigen expression in prostatic intraepithelial neoplasia and adenocarcinoma: a study of 184 cases. Cancer 82:2256–2261
51. Pillai MRA, Nanabala R, Joy A, Sasikumar A, Knapp FF (2016) Radiolabeled enzyme inhibitors and binding agents targeting PSMA: effective theranostic tools for imaging and therapy of prostate cancer. Nucl Med Biol 43:692–720
52. Afshar-Oromieh A, Haberkorn U, Eder M, Eisenhut M, Zechmann CM (2012) [^{68}Ga]gallium-labelled PSMA ligand as superior PET tracer for the diagnosis of prostate cancer: comparison with ^{18}F-FECH. Eur J Nucl Med Mol Imaging 39:1085–1086
53. Fendler WP et al (2017) Establishing ^{177}Lu-PSMA-617 radioligand therapy in a syngeneic model of murine prostate cancer. J Nucl Med 58:1786–1792
54. Afshar-Oromieh A et al (2013) PET imaging with a [^{68}Ga]gallium-labelled PSMA ligand for the diagnosis of prostate cancer: biodistribution in humans and first evaluation of tumour lesions. Eur J Nucl Med Mol Imaging 40:486–495
55. Budäus L et al (2016) Initial experience of 68Ga-PSMA PET/CT imaging in high-risk prostate cancer patients prior to radical prostatectomy. Eur Urol 69:393–396
56. Thomas L, Balmus C, Ahmadzadehfar H, Essler M, Strunk H, Bundschuh RA (2017) Assessment of bone metastases in patients with prostate cancer—a comparison between 99mTc-bone-scintigraphy and [^{68}Ga]Ga-PSMA PET/CT. Pharmaceuticals 10:68. https://doi.org/10.3390/ph10030068
57. Markwalder R, Reubi JC (1999) Gastrin-releasing peptide receptors in the human prostate: relation to neoplastic transformation. Cancer Res 59:1152–1159
58. Pinski J, Halmos G, Yano T, Szepeshazi K, Qin Y, Ertl T, Schally AV (1994) Inhibition of growth of MKN45 human gastric-carcinoma xenografts in nude mice by treatment with bombesin/gastrin-releasing-peptide antagonist (RC-3095) and somatostatin analogue RC-160. Int J Cancer 57:574–580
59. Sun B, Schally AV, Halmos G (2000) The presence of receptors for bombesin/GRP and mRNA for three receptor subtypes in human ovarian epithelial cancers. Regul Pept 90:77–84
60. Cescato R, Maina T, Nock B, Nikolopoulou A, Charalambidis D, Piccand V, Reubi JC (2008) Bombesin receptor antagonists may be preferable to agonists for tumor targeting. J Nucl Med 49:318–326
61. Siegfried JM, Krishnamachary N, Gaither Davis A, Gubish C, Hunt JD, Shriver SP (1999) Evidence for autocrine actions of neuromedin B and gastrin-releasing peptide in non-small cell lung cancer. Pulm Pharmacol Ther 12:291–302
62. Liu Y, Karaca M, Zhang Z, Gioeli D, Earp HS, Whang YE (2010) Dasatinib inhibits site-specific tyrosine phosphorylation of androgen receptor by Ack1 and Src kinases. Oncogene 29:3208–3216
63. Wen X, Chao C, Ives K, Hellmich MR (2011) Regulation of bombesin-stimulated cyclo-oxygenase-2 expression in prostate cancer cells. BMC Mol Biol 12:29. https://doi.org/10.1186/1471-2199-12-29

64. Reubi JC, Fleischmann A, Waser B, Rehmann R (2011) Concomitant vascular GRP-receptor and VEGF-receptor expression in human tumors: molecular basis for dual targeting of tumoral vasculature. Peptides 32:1457–1462
65. Jensen RT, Battey JF, Spindel ER, Benya RV (2008) International Union of Pharmacology. LXVIII. Mammalian bombesin receptors: nomenclature, distribution, pharmacology, signaling, and functions in normal and disease states. Pharmacol Rev 60:1–42
66. Ananias HJ, van den Heuvel MC, Helfrich W, de Jong IJ (2009) Expression of the gastrin-releasing peptide receptor, the prostate stem cell antigen and the prostate-specific membrane antigen in lymph node and bone metastases of prostate cancer. Prostate 69:1101–1108
67. Liu Z, Niu G, Wang F, Chen X (2009) ^{68}Ga-labeled NOTA-RGD-BBN peptide for dual integrin and GRPR-targeted tumor imaging. Eur J Nucl Med and Mol Imag 36:1483–1494
68. Anastasi A, Erspamer V, Bucci M (1972) Isolation and amino acid sequences of alytesin and bombesin, two analogous active tetradecapeptides from the skin of European discoglossid frogs. Arch Biochem Biophys 148:443–446
69. Baratto L, Jadvar H, Iagaru A (2017) Prostate cancer theranostics targeting gastrin-releasing peptide receptors. Mol Imaging Biol. https://doi.org/10.1007/s11307-0171151-1
70. Maddalena ME et al (2009) ^{177}Lu-AMBA biodistribution, radiotherapeutic efficacy, imaging, and autoradiography in prostate cancer models with low GRP-R expression. J Nucl Med 50:2017–2024
71. Maina T, Nock B, Mather S (2006) Targeting prostate cancer with radiolabelled bombesins. Cancer Imaging 6:153–157
72. Nock BA, Nikolopoulou A, Galanis A, Cordopatis P, Waser B, Reubi JC, Maina T (2005) Potent bombesin-like peptides for GRP-receptor targeting of tumors with ^{99m}Tc: a preclinical study. J Med Chem 48:100–110
73. Yu Z et al (2013) An update of radiolabeled bombesin analogs for gastrin-releasing peptide receptor targeting. Curr Pharm Des 19:3329–3341
74. Zhang H, Schuhmacher J, Waser B, Wild D, Eisenhut M, Reubi JC, Maecke HR (2007) DOTA-PESIN, a DOTA-conjugated bombesin derivative designed for the imaging and targeted radionuclide treatment of bombesin receptor-positive tumours. Eur J Nucl Med Mol Imaging 34:1198–1208
75. Scopinaro F et al (2003) ^{99m}Tc-bombesin detects prostate cancer and invasion of pelvic lymph nodes. Eur J Nucl Med Mol Imaging 30:1378–1382
76. Van de Wiele C et al (2000) Technetium-99m RP527, a GRP analogue for visualisation of GRP receptor-expressing malignancies: a feasibility study. Eur J Nucl Med 27:1694–1699
77. Dijkgraaf I et al (2012) PET of tumors expressing gastrin-releasing peptide receptor with an ^{18}F-labeled bombesin analog. J Nucl Med 53:947–952
78. Carlucci G et al (2015) GRPR-selective PET imaging of prostate cancer using [(^{18}F)]-lanthionine-bombesin analogs. Peptides 67:45–54
79. Lane SR et al (2010) Optimization, biological evaluation and microPET imaging of ^{64}Cu-labeled bombesin agonists, [^{64}Cu-NO2A-(X)-BBN(7-14)NH2], in a prostate tumor xenografted mouse model. Nucl Med Biol 37:751–761
80. Bass LA, Wang M, Welch MJ, Anderson CJ (2000) In vivo transchelation of ^{64}Cu from TETA-octreotide to superoxide dismutase in rat liver. Bioconjug Chem 11:527–532
81. Garrison JC, Rold TL, Sieckman GL, Figueroa SD, Volkert WA, Jurisson SS, Hoffman TJ (2007) In vivo evaluation and small-animal PET/CT of a prostate cancer mouse model using ^{64}Cu bombesin analogs: side-by-side comparison of the CB-TE2A and DOTA chelation systems. J Nucl Med 48:1327–1337
82. Shokeen M, Anderson CJ (2009) Molecular imaging of cancer with ^{64}Cu radiopharmaceuticals and positron emission tomography (PET). Acc Chem Res 42:832–841
83. Mansi R et al (2009) Evaluation of a 1,4,7,10-tetraazacyclododecane-1,4,7,10-tetraacetic acid-conjugated bombesin-based radioantagonist for the labeling with single-photon emission computed tomography, positron emission tomography, and therapeutic radionuclides. Clin Cancer Res 15:5240–5249

84. Mansi R et al (2011) Development of a potent DOTA-conjugated bombesin antagonist for targeting GRPr-positive tumours. Eur J Nucl Med Mol Imaging 38:97–107
85. Gourni E et al (2014) N-terminal modifications improve the receptor affinity and pharmacokinetics of radiolabeled peptidic gastrin-releasing peptide receptor antagonists: examples of ^{68}Ga- and ^{64}Cu-labeled peptides for PET imaging. J Nucl Med 55:1719–1725
86. Dalm SU et al (2017) ^{68}Ga/^{177}Lu-NeoBOMB1, a novel radiolabeled GRPR antagonist for theranostic use in oncology. J Nucl Med 58:293–299
87. Chatalic KL et al (2014) Preclinical comparison of Al^{18}F- and ^{68}Ga-labeled gastrin-releasing peptide receptor antagonists for PET imaging of prostate cancer. J Nucl Med 55:2050–2056
88. Pan D et al (2014a) A new ^{68}Ga-labeled BBN peptide with a hydrophilic linker for GRPR-targeted tumor imaging. Amino Acids 46:1481–1489
89. Pan D et al (2014b) PET imaging of prostate tumors with ^{18}F-Al-NOTA-MATBBN contrast media. Mol Imaging 9:342–348
90. Yang M et al (2011) ^{18}F-labeled GRPR agonists and antagonists: a comparative study in prostate cancer imaging. Theranostics 1:220–229
91. Kahkonin E et al (2013) In vivo imaging of prostate cancer using [^{68}Ga]-labeled bombesin analog BAY86-7548. Clin Cancer Res 19:5434–5443
92. Wieser G et al (2014) Positron emission tomography (PET) imaging of prostate cancer with a gastrin releasing peptide receptor antagonist – from mice to men. Theranostics 4:412–419
93. Wieser G et al (2017) Diagnosis of recurrent prostate cancer with PET/CT imaging using the gastrin-releasing peptide receptor antagonist ^{68}Ga-RM2: preliminary results in patients with negative or inconclusive [^{18}F]fluoroethylcholine-PET/CT. Eur J Nucl Med Mol Imaging 44:1463–1472
94. Minamimoto R et al (2016) Pilot comparison of ^{68}Ga-RM2 PET and ^{68}Ga-PMSA-11 PET in patients with biochemically recurrent prostate cancer. J Nucl Med 57:557–562
95. Minamimoto R, Sonni I, Hancock S, Vasanawala S, Loening A, Gambhir SS, Iagaru A (2017) Prospective evaluation of ^{68}Ga-RM2 PET/MRI in patients with biochemical recurrence of prostate cancer and negative conventional imaging. J Nucl Med 59:803–808. https://doi.org/10.2967/jnumed.117.197624
96. Romanov VI, Goligorsky MS (1999) RGD-recognizing integrins mediate interactions of human prostate carcinoma cells with endothelial cells in vitro. Prostate 39:108–118
97. Ruoslahti E, Pierschbacher MD (1987) New perspectives in cell adhesion: RGD and integrins. Science 238:491–497
98. Sutherland M, Gordon A, Shnyder SD, Patterson LH, Sheldrake HM (2012) RGD-binding integrins in prostate cancer: expression patterns and therapeutic prospects against bone metastasis. Cancers (Basel) 4:1106–1145
99. Christofori G (2003) Changing neighbours, changing behaviour: cell adhesion molecule-mediated signalling during tumour progression. EMBO J 22:2318–2323
100. Cooper CR, Chay CH, Pienta KJ (2002) The role of alpha(v)beta(3) in prostate cancer progression. Neoplasia 4:191–194
101. Haass NK, Smalley KS, Li L, Herlyn M (2005) Adhesion, migration and communication in melanocytes and melanoma. Pigment Cell Res 18:150–159
102. Hood JD, Cheresh DA (2002) Role of integrins in cell invasion and migration. Nat Rev Cancer 2:91–100
103. Hynes RO (2002) Integrins: bidirectional, allosteric signaling machines. Cell 110:673–687
104. Pavalko FM, Otey CA (1994) Role of adhesion molecule cytoplasmic domains in mediating interactions with the cytoskeleton. Proc Soc Exp Biol Med 205:282–293
105. Jackson AB et al (2012) ^{64}Cu-NO2A-RGD-Glu-6-Ahx-BBN(7-14)NH2: a heterodimeric targeting vector for positron emission tomography imaging of prostate cancer. Nucl Med Biol 39:377–387
106. Shallal HM, Minn I, Banerjee SR, Lisok A, Mease RC, Pomper MG (2014) Heterobivalent agents targeting PSMA and integrin-$\alpha v\beta 3$. Bioconjug Chem 25:393–405

107. Taylor RM, Severns V, Brown DC, Bisoffi M, Sillerud LO (2012) Prostate cancer targeting motifs: expression of $\alpha v\beta 3$, neurotensin receptor 1, prostate specific membrane antigen, and prostate stem cell antigen in human prostate cancer cell lines and xenografts. Prostate 72:523–532
108. Xiong JP, Stehle T, Zhang R, Joachimiak A, Frech M, Goodman SL, Arnaout MA (2002) Crystal structure of the extracellular segment of integrin $\alpha v\beta 3$ in complex with an Arg-Gly-asp ligand. Science 296:151–155
109. Liu S (2006) Radiolabeled multimeric cyclic RGD peptides as integrin alphavbeta3 targeted radiotracers for tumor imaging. Mol Pharm 3:472–487
110. Shi J, Wang F, Liu S (2016) Radiolabeled cyclic RGD peptides as radiotracers for tumor imaging. Biophys Rep 2:1–20
111. Arap W, Pasqualini R, Ruoslahti E (1998) Cancer treatment by targeted drug delivery to tumor vasculature in a mouse model. Science 279:377–380
112. Chen X et al (2004) ^{18}F-labeled RGD peptide: initial evaluation for imaging brain tumor angiogenesis. Nucl Med Biol 31:179–189
113. Chen X, Plasencia C, Hou Y, Neamati N (2005) Synthesis and biological evaluation of dimeric RGD peptide-paclitaxel conjugate as a model for integrin-targeted drug delivery. J Med Chem 48:1098–1106
114. Kim JW, Lee HS (2004) Tumor targeting by doxorubicin-RGD-4C peptide conjugate in an orthotopic mouse hepatoma model. Int J Mol Med 14:529–535
115. Wu Y et al (2005) microPET imaging of glioma integrin {alpha}v{beta}3 expression using ^{64}Cu-labeled tetrameric RGD peptide. J Nucl Med 46:1707–1718
116. Janssen M et al (2002) Comparison of a monomeric and dimeric radiolabeled RGD-peptide for tumor targeting. Cancer Biother Radiopharm 17:641–646
117. Schottelius M, Laufer B, Kessler H, Wester HJ (2009) Ligands for mapping alphavbeta3-integrin expression in vivo. Acc Chem Res 42:969–980
118. Beer AJ et al (2008) Patterns of alphavbeta3 expression in primary and metastatic human breast cancer as shown by ^{18}F-Galacto-RGD PET. J Nucl Med 49:255–259
119. Blom E et al (2012) (68)Ga-labeling of RGD peptides and biodistribution. Int J Clin Exp Med 5:165–172
120. Dijkgraaf I, Kruijtzer JA, Frielink C, Corstens FH, Oyen WJ, Liskamp RM, Boerman OC (2007) Alpha v beta 3 integrin-targeting of intraperitoneally growing tumors with a radiolabeled RGD peptide. Int J Cancer 120:605–610
121. Dijkgraaf I et al (2013) Imaging integrin alpha-v-beta-3 expression in tumors with an ^{18}F-labeled dimeric RGD peptide. Contrast Media Mol Imaging 8:238–245
122. Li ZB, Cai W, Cao Q, Chen K, Wu Z, He L, Chen X (2007) ^{64}Cu-labeled tetrameric and octameric RGD peptides for small-animal PET of tumor alpha(v)beta(3) integrin expression. J Nucl Med 48:1162–1171
123. Wu Z et al (2007) microPET of tumor integrin alphavbeta3 expression using ^{18}F-labeled PEGylated tetrameric RGD peptide (18F-FPRGD4). J Nucl Med 48:1536–1544
124. Andriu A, Crockett J, Dall'Angelo S, Piras M, Zanda M, Fleming IN (2018) Binding of $\alpha_V\beta_3$ integrin-specific radiotracers is modulated by both integrin expression level and activation status. Mol Imaging Biol 20:27–36
125. Israel I, Richter D, Stritzker J, van Ooschot M, Donat U, Buck AK, Samnick S (2014) PET imaging with [^{68}Ga]NOTA-RGD for prostate cancer: a comparative study with [^{18}F]fluorode-oxyglucose and [^{18}F]fluoroethylcholine. Curr Cancer Drug Targets 14:371–379
126. Hu K et al (2015) ^{18}F-FP-PEG2-beta-Glu-RGD2: a symmetric integrin $\alpha_v\beta_3$-targeting radiotracer for tumor PET imaging. PLoS One 10:e0138675
127. Cheng Z et al (2015) Ex-vivo biodistribution and micro-PET/CT imaging of ^{18}F-FDG, ^{18}F-FLT, ^{18}F-FMISO, and ^{18}F-AlF-NOTA-PRGD2 in a prostate tumor-bearing nude mouse model. Nucl Med Commun 36:914–921
128. Beer AJ et al (2016) Non-invasive assessment of inter- and intrapatient variability of integrin expression in metastasized prostate cancer by PET. Oncotarget 7:28151–28159

129. Lantry LE et al (2006) ¹⁷⁷Lu-AMBA: aynthesis and characterization of a selective 177Lu-labeled GRP-R agonist for systemic radiotherapy of prostate cancer. J Nucl Med 47:1144–1152

130. Lantry LE et al (2004) Preclinical evaluation of ¹⁷⁷Lu-AMBA, a DOTA conjugate that targets GRP and NMB receptor expressing tumors: internalization, in vivo biodistribution, single dose radiotherapy in PC-3 tumor-bearing nude mice and in vitro autoradiography in animal and human tissues. EANM, Helsinki

131. Bodei L et al (2007) ¹⁷⁷Lu-AMBA Bombesin analogue in hormone refractory prostate cancer patients: a phase I escalation study with single-cycle administrations. In: Annual congress, European Association of Nuclear Medicine, Copenhagen, Denmark, 13–17 Oct 2007

132. Li ZB, Wu Z, Chen K, Ryu EK, Chen X (2008) ¹⁸F-labeled BBN-RGD heterodimer for prostate cancer imaging. J Nucl Med 49:453–461

133. Zhang J et al (2017) Clinical translation of a dual integrin $\alpha_v\beta_3$- and gastrin-releasing peptide receptor-targeting PET radiotracer, ⁶⁸Ga-BBN-RGD. J Nucl Med 58:228–234

134. Durkan K et al (2014) A heterodimeric [RGD-Glu-[⁶⁴Cu-NO2A]-6-Ahx-RM2] $\alpha_v\beta_3$/GRPr-targeting antagonist radiotracer for PET imaging of prostate tumors. Nucl Med Biol 41:133–139

135. Stott Reynolds TJ et al (2015) Characterization and evaluation of DOTA-conjugated Bombesin/RGD-antagonists for prostate cancer tumor imaging and therapy. Nucl Med Biol 42:99–108

Chapter 9
Targeted Prostate Biopsy and MR-Guided Therapy for Prostate Cancer

David A. Woodrum, Akira Kawashima, Krzysztof R. Gorny, and Lance A. Mynderse

Abstract In 2018, the American Cancer Society (ACS) estimates that 164,690 new cases of prostate cancer will be diagnosed and 29,430 will die due to the prostate cancer in the United States (Siegel et al., CA Cancer J Clin 67:7–30, 2018). Many men with prostate cancer are often managed with aggressive therapy including radiotherapy or surgery. No matter how expertly done, these therapies carry significant risk and morbidity to the patient's health related quality of life with impact on sexual, urinary and bowel function (Potosky et al., J Natl Cancer Inst 96:1358–1367, 2004). A recent meta-analysis of 19 studies reviewing the use of surgery and radiation for prostate cancer demonstrated patients who received radiation were more likely to die from their disease as compared to surgery (Wallis et al., Eur Urol 70:21–30, 2016). Furthermore, screening programs using prostatic specific antigen (PSA) and transrectal ultrasound (TRUS) guided systematic biopsy have identified increasing numbers of low risk, low grade "localized" prostate cancer. This indolent nature of many prostate cancers presents a difficult decision of when to intervene given the possible comorbidities of aggressive treatment. Active surveillance has been increasingly instituted in order to balance cancer control versus treatment side effects (Jemal et al., CA Cancer J Clin 56:106–130, 2006). Although active debate continues on the suitability of focal or regional therapy for these low or intermediate risk prostate cancer patients, many unresolved issues remain which complicate this approach of management. Some of the largest unresolved issues are: prostate cancer multifocality, limitations of current biopsy strategies, suboptimal staging by accepted imaging modalities, less than

D. A. Woodrum (✉) · K. R. Gorny
Department of Radiology, Mayo Clinic, Rochester, MN, USA
e-mail: woodrum.david@mayo.edu; gorny.krzysztof@mayo.edu

A. Kawashima
Department of Radiology, Mayo Clinic, Scottsdale, AZ, USA
e-mail: kawashima.akira@mayo.edu

L. A. Mynderse
Department of Urology, Mayo Clinic, Rochester, MN, USA
e-mail: mynderse.lance@mayo.edu

© Springer Nature Switzerland AG 2018
H. Schatten (ed.), *Molecular & Diagnostic Imaging in Prostate Cancer*,
Advances in Experimental Medicine and Biology 1096,
https://doi.org/10.1007/978-3-319-99286-0_9

159

robust prediction models for indolent prostate cancers and whether established curative therapies can be safely and effectively used following focal therapy for prostate cancer. In spite of these restrictions focal therapy continues to confront the current paradigm of therapy for low and even intermediate risk disease (Onik, Tech Vasc Interv Radiol 10:149–158, 2017). It has been proposed that early detection and proper characterization may play a role in preventing the development of metastatic disease (Vickers et al., BMJ 346:f2023, 2013). There is Level 1 evidence supporting detection and subsequent aggressive treatment of intermediate and high-risk prostate cancer (Bill-Axelson et al., N Engl J Med 370:932–942, 2014). Therefore accurate assessment of cancer risk (i.e. grade and stage) using imaging and targeted biopsy is critical. Advances in prostate imaging with MRI have been accompanied with advances in MR guided therapy propelling prostate treatment solutions forward faster than ever.

9.1 Primary Prostate Cancer

9.1.1 Cancer Work-Up

The historical workup for prostate cancer has been a combination of prostate specific antigen (PSA) screening and digital rectal exam (DRE) followed by DRE directed biopsy. More recently, the use of ultrasound (US) imaging has helped to direct the biopsies toward suspicious lesions and systematically sample the prostate. US alone is not sensitive enough to find all the prostate cancer within the gland in spite of advanced US modalities (i.e. color/power doppler, elastography and bubble contrast agents). Furthermore, systematic (non-targeted) sampling the entire organ has provided some answers but also runs the risk of under sampling small volume high grade but clinically significant disease or over sampling indolent low grade disease, potentially resulting in delayed diagnosis and over treatment.

Magnetic resonance imaging (MRI) is the superior imaging modality for prostate and associated structures due to exceptional soft tissue conspicuity, high spatial resolution, and cross-sectional imaging. Utilization of integrated endorectal and pelvic phased-array coils has led to continued improvement in prostatic fossa visualization. High resolution T2-weighted imaging is sensitive in depicting prostate cancer especially within the transition zone of the prostate. However, decreased T2 signal intensity is not specific for prostate cancer especially within the peripheral zone where benign conditions can lead to imaging changes. Functional parametric imaging including dynamic contrast-enhanced imaging (DCEI), diffusion-weighted imaging (DWI), and MR spectroscopic imaging (MRSI) complement morphologic MRI by reflecting perfusion characteristics, Brownian motion of water molecules, and metabolic profiles, respectively. Significant inverse correlation was shown between ADC value and Gleason score/highest grade [4].

In 2013, a consensus panel endorsed utilization of mpMRI to identify patients for focal therapy [5]. Multiparametric MRI is capable of localizing small tumors for focal therapy. In 2015, a consensus panel agreed to Prostate Imaging Reporting and Data System (PI-RADS) version 2 which promoted standardized MR acquisition and interpretation to improve detection, localization, characterization, and risk stratification of clinically significant prostate cancer in treatment naïve prostate glands [6]. Targeted biopsy of suspected cancer lesions detected by MRI is associated with increased detection of high-risk prostate cancer and decreased detection of low-risk prostate cancer particularly with the aid of MRI/Ultrasound fusion platforms [7]. The use of mpMRI has expanded beyond staging to detection, characterization, monitoring for active surveillance, and cases of suspected local recurrence after failed definitive therapy.

While mpMRI plays an established, critical role in primary and recurrent prostate cancer, functional and metabolic imaging, are playing an expanding role with a host of new agents being developed. Established positron emission tomography (PET) tracers for imaging of prostate cancer include ^{11}C and ^{18}F choline, ^{18}F fluciclovine, ^{68}Ga prostate specific membrane antigen (PSMA), and ^{11}C-acetate. ^{11}C-choline PET/CT has an advantage to reveal both local recurrent and distant metastatic prostate cancers. ^{11}C-choline PET/CT had a sensitivity of 73%, a specificity of 88%, a positive predictive value (PPV) of 92%, a negative predictive value (NPV) of 61%, and an accuracy of 78% for the detection of clinically suspected recurrent prostate cancer in postsurgical patients [8]. In a study of post-prostatectomy patients with rising PSA, mpMRI is superior for the detection of local recurrence, ^{11}C-choline PET/CT is superior for pelvic nodal metastasis, and both are equally excellent for pelvic bone metastasis. ^{11}C-choline PET/CT and mp-MR imaging are complementary for restaging prostatectomy patients with suspected recurrent disease and exhibit diverse patterns of recurrence with implications for optimal salvage treatment strategies [9, 10]. ^{68}Ga-PSMA is a promising PET tracer and indicates favorable sensitivity and specificity profiles compared to choline-based PET imaging techniques [11]. A recent publication demonstrated that late 3 h imaging of ^{68}Ga-PSMA helped to clarify activity within the prostate due to decreased activity within the bladder at this time point [12]. Early work with simultaneous MRI/PET imaging shows promise in capitalizing both the functional aspects of PET with the superb anatomic capabilities of MRI [13].

Even with improvements in US and PET/CT imaging, MRI remains preeminent for detection and staging of prostate tumors within the pelvis. MRI/PET may ultimately provide the optimal combination of diagnostic resolution in the pelvis coupled with the whole-body screening functionality of PET imaging to provide the single platform for detection and localization.

9.2   Biopsy Methods

9.2.1   Prostate Biopsy Techniques

9.2.1.1   Ultrasound Guided Biopsies

The TRUS prostate biopsy has remained the cornerstone for prostate cancer tissue diagnosis dating back to the systematic 'sextant' biopsy protocol with three cores per side [14]. A meta-analysis of 68 studies led to a recommendation of a more laterally directed schema with 12 cores improving prostate cancer detection rates by a factor of 1.3 [15]. Using this systematic 12 core TRUS sampling for men undergoing initial biopsy with elevated PSA yields cancer detection rates between 30% and 55% [16]. The false negative rate for this 12 core schema is on the order of 20–24% [17] and repeated 12 core or saturation biopsies show detection rates of 11–47% [18]. This is particularly true for men with anteriorly located and apex tumors [19]. To improve the accuracy of the sampling, some experts advocate the use of template, transperineal-mapping biopsies to systematically sample all quadrants of the prostate [20]. This has been criticized for oversampling of insignificant tumors with risk of additional morbidity and need for general anesthesia.

9.2.1.2   MR-Based Biopsy Techniques

Increasingly, evidence supports the use of pre-biopsy mpMRI for identification of clinically significant disease. The hope is to identify the significant lesions for targeted biopsy while not oversampling otherwise normal regions [20–22]. There are three main MR-based biopsy technical approaches.

Cognitive/Visual-Directed MRI Targeted Biopsy

Overall, cognitive fusion techniques demonstrate the most variability between operators due to the reliance on spatial orientation of the lesion from the MR which is used to direct the US and biopsy needle by the operator. With appropriate experience, this can be readily implemented but can be very difficult with small lesions or targets well away from the US transducer such as with anterior tumors in large glands. Although MR-directed cognitive fusion biopsy of the prostatectomy bed is still useful for the evaluation of patients with biochemical failure after surgery, no dedicated MR-TRUS fusion software biopsy system is currently available for this application. The cognitive/fusion method is prone to error in reliably mapping the MRI suspicious lesion on real time TRUS and the confirmation of TRUS guided targeted biopsy needle location over the MRI suspicious lesion is not feasible except when needle tracks are visible. In a study of 555 patients by systematic biopsy as well as cognitive fusion guided targeted biopsy, overall 54% (302/555) of patients were found to have cancer; 82% of them were clinically significant. Systematic biopsy and cognitive fusion guided targeted biopsy detected 88% and 98% of clinically significant cancers, respectively [23]. Cognitive fusion targeted biopsy showed 16% more high-grade cancers and higher mean cancer core lengths than standard systematic biopsy. Cognitive targeted biopsy would only avoid 13% of insignificant tumors. Valerio et al. compared cognitive fusion to a software-based targeted biopsy. The software-based, targeted transperineal approach found more clinically significant disease than visually directed biopsy although this was not statistically significant (51.9% vs. 44.3%, p = 0.24) [24]. The current diagnostic ability of visually/cognitively targeted and software-based biopsies seem to be nearly comparable with experienced operators.

Software Based Ultrasound–MRI Fusion Targeted Biopsy

Software based MRI/TRUS fusion guided biopsy platforms seek to combine the advantage of lesion visualization from the MRI with the ease and availability of US based biopsy platforms for real-time imaging. There are three key tracking methods including (1) image organ-based tracking, (2) electromagnetic sensor-based tracking, and (3) mechanical arm, sensor-based tracking [25].

Image organ (prostate)-based tracking method fuses prior MRI with real time 3D US using a surface-based registration and elastic organ-based deformation algorithm (Urostation, Koelis, Meylan, France). MRI-identified suspicious lesions are loaded into the system which then projects the target into the biopsy aiming mechanism on the US probe. This is relatively inexpensive and allows systematic biopsy. However, confirmation of targeted needle biopsy tracts is retrospective [26].

Additional systems utilize electromagnetic sensor-based tracking using a non-rigid registration algorithm. The advantage with this approach is that it allows real-time spatial tracking of targets and needle location and is less operator-dependent allowing free hand scanning during procedures (UroNav, InVivo, Inc., Gainsville, FL, USA; Real-time Virtual Sonography [RVS], Hitachi-Aloka, Tokyo, Japan and BK Fusion, BK Medical ApS, Herlev, Denmark). In a recent prospective study of 1003 men undergoing a MR/ultrasound fusion targeted biopsy and concurrent standard biopsy, targeted MR/ultrasound fusion biopsy was shown to diagnose 30% more high-risk prostate cancer (defined as Gleason score 4 + 3 or greater) while a combination of standard and targeted biopsies revealed 22% more prostate cancer, mostly (83%) low-risk prostate cancer (defined as Gleason score 3 + 3 and low volume 3 + 4) [27].

A similar but slightly different approach uses a mechanical arm, sensor-based tracking system where the tracking arm is attached to a conventional US probe. Again this allows real-time spatial tracking of targets and needle location (Artemis, Eigen Inc., Grass Valley, CA, USA). This system is also less operator-dependent but relatively expensive. In a recent retrospective review of 601 men who underwent both MRI-ultrasound fusion targeted biopsy and systematic biopsy, targeted MRI-ultrasound fusion biopsy detected fewer Gleason score 6 prostate cancers (75 vs.

121; p < 0.001) and more Gleason score ≥ 7 prostate cancers (158 vs. 117; p < 0.001) when compared with systemic biopsy [28]. In a review of 105 patients with prior negative biopsies and elevated PSA, MRI-ultrasound fusion targeted biopsy improved detection of clinically significant prostate cancer when compared with systemic biopsy [29].

Recent advances in MRI have demonstrated the value and importance of good prostate imaging. These advances are actively changing the way prostate cancer is diagnosed and treated. However, even with these advances it is critical to understand that MRI still has its limitations and needs further development. A recent study of 125 surgical prostatectomy patients studied the accuracy of the pre-surgical biopsy where there had been a pre-biopsy mpMRI with subsequent MR-US fusion biopsy. They found that there was a 4% (5 of 123) MR miss rate on surgical pathologic analysis [30]. Another study of 1042 men examined mpMRI targeted biopsies versus systematic biopsies. They found that the addition of systematic biopsy to targeted biopsy found 7% (60/825) more clinically significant cancers [31]. These lesions would have been underdiagnosed if mpMR suspected lesions only were targeted. In a recent sturdy in 100 patients who underwent mpMRI, 162 clinically important malignant lesions were present after subsequent prostatectomy. On a per patient basis, mpMRI depicted clinically important prostate cancer in 99 (99%) of the 100 patients. However, at least one clinically important tumor was missed in 26 (26%) patients [32].

In-Bore Direct MRI Targeted Biopsy

There are two main in-bore direct MRI targeted biopsy approaches including robot assisted, transrectal biopsy (DynaTRIM, InVivo) or a transperineal approach via template. In-bore MRI-guided biopsies have the advantage of realtime MR imaging to confirm biopsy acquisition position. Using the direct in-bore biopsy technique one eliminates the issues of mis-registration, organ deformation and organ movement which continue to plague software based US fusion imaging. Additionally, transperineal biopsies using a template reduce or nearly eliminate the bacterial infection risk seen in trans-rectal biopsies. A study of 265 patients with rising PSA and negative TRUS biopsy found that performance of MR guided robot assisted transrectal biopsy (DynaTRIM, InVivo) in this population produced a detection rate of 41% (108/265) for prostate cancer and 87% (94/108) for clinically significant cancer [29]. Penszkofer et al. showed in a prospective clinical study, that in-bore prostate biopsies with at least one MRI detected lesion in men on active surveillance monitoring and in men with suspected recurrent cancer following treatment, detected cancer in 72% under active surveillance and detected recurrent cancer in 72% with possible recurrence [33, 34]. The accuracy of the transperineal in-bore biopsy appears acceptable as demonstrated by analysis of biopsy and post prostatectomy histopathology [35]. MRI detected targets located in the anterior gland had the highest cancer yield (62.5%). Although these advantages are attractive, this technique is underutilized due to specialized MRI compatible tools, relative cost disadvantage, difficulty obtaining access to MRI, and need for coordination between Urology and Radiology.

9.2.1.3   Treatments for Primary Prostate Cancer

Once prostate cancer is identified from imaging and/or biopsy a treatment plan must be formulated for the patient. The traditional therapy options for clinically localized prostate cancer with intent for cure have been either surgical resection or radiotherapy [36]. A recent meta-analysis of 19 studies suggests that surgery offers a benefit in overall and prostate cancer-specific survival compared with radiotherapy [2]. For patients with localized high-risk prostate cancer, recent reviews suggest a benefit in radical prostatectomy over radiotherapy for overall and prostate cancer specific mortality [37, 38]. Roughly about half of patients choose surgery and half choose radiotherapy [39].

However, these therapies have significant risk and morbidity to the patient's health related quality of life with potential impact on sexual, urinary and bowel function [1]. Active screening programs for prostate cancer have enabled earlier identification of low risk prostate cancer, but due to related morbidity from standard therapies, many choose active surveillance to delay treatment until cancer progression [3].

9.2.1.4   Evolving Focal and Partial Gland Therapy Treatment Options

For men with newly diagnosed prostate cancer and with a life expectancy >10 years, radical prostatectomy and radiation therapy remain preferred definitive therapy of choice [40–42]. However, patients are increasingly interested in less radical, more focal, methodologies for treatment especially in the active surveillance population. For this population of low- and intermediate-risk prostate cancer patients, they also may be uncomfortable remaining on active surveillance but don't want surgery or radiation. This patient-driven interest for a more minimally invasive approach is driving focal therapies for prostate carcinoma in low-risk patients [43]. As a result, several minimally invasive thermal ablation methods under direct MR guidance, most prominently cryotherapy [44], laser ablation [45], and high-intensity focused ultrasound (HIFU) [46], have been developed and are currently being evaluated. Despite this, focal therapy is still controversial due to the potential for multifocality of prostate cancer, limitations of current biopsy strategies, variation in quality MR imaging and less than robust prediction models for indolent prostate cancers. Furthermore, prostate cancer recurrence rates after established forms of therapy range from 20% to 60% [47].

Patient Selection for Focal and Partial Gland Therapy

Selecting the appropriate patient for focal/partial gland therapy as a primary treatment for prostate cancer is the most important element of a successful outcome. Accurately staging the prostate cancer is critical not only for the highest grade of prostate cancer but also understanding the extent of low-grade (i.e. Gleason 6) disease as well. With low-risk disease, there is level 1 evidence that implies a lack of benefit from radical/nonconservative therapy [48–50]. Patients are many times initially targeted for cancer workup due to rising PSA or nodule on digital rectal exam. These patients may undergo further workup with mapping biopsy and/or mpMRI with targeted biopsy. Patients are then classified to have low, intermediate, or high risk disease. For consideration of focal therapy, the patient needs to have low or intermediate risk disease with a focal positive index lesion on mpMRI, Gleason $\leq 4 + 3$, and PSA < 20 ng/mL. The target lesion should be confined to one lobe of the prostate [49]. Furthermore, the target should be visible with the imaging modality which will be used to guide the focal ablation treatment. With MRI having exceptional soft tissue conspicuity, high spatial resolution, and multi-planar imaging capacity, this approach has clear advantage over transrectal ultrasound as an imaging guidance platform.

Modalities Used to Ablate Primary Prostate Cancer: MR-Guided Cryoablation

MR guided percutaneous cryoablation combines excellent soft tissue resolution and ice ball monitoring without the use of MRI thermometry. Early experience combining cryoablation with MRI has shown a high degree of accuracy in defining normal and frozen tissue on all MR imaging sequences [51, 52]. There is limited data using MR guided cryoablation for primary treatment of prostate cancer. Two published canine studies demonstrated feasibility and overall safety [53, 54]. These studies did expose one limitation of cryoablation which is that the visualized edge of the ice (0 °C) does not represent the ablation margin. The actual ablation margin is best demonstrated with contrast enhancement post-procedure and is actually at the −20 °C isotherm. There are two published reports of MR guided cryoablation in native prostate glands, each with relatively small numbers (Fig. 9.1) [55, 56]. Gangi et al. performed MRI guided prostate cryoablation in eleven patients on 1.5 T MRI with minor complications of hematuria, dysuria, and urine retention and one major complication of rectal fistula with spontaneous closure after 3 months [56]. These studies confirm that MR guided cryoablation is technically feasible with relative safety, however, more intermediate and long term outcome data is needed to assess overall efficacy.

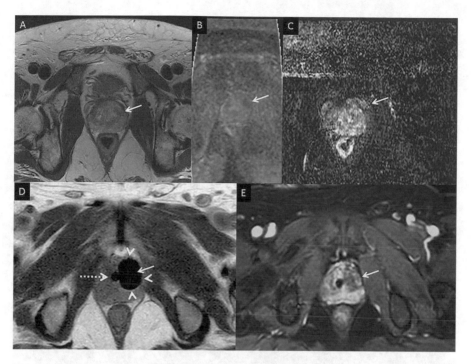

Fig. 9.1 Seventy-year-old male presents with Gleason 3 + 3 adenocarcinoma of the left anterior prostate with a PSA of 6.2 ng/mL. Multiparametric MRI at 3 T demonstrated an area of decreased signal intensity in the left anterolateral peripheral zone at the prostatic midgland (Panel **a**, arrow) with corresponding hypointensity on ADC map (Panel **b**, arrow), and early hyperenhancement (Panel **c**, arrow). Using a 1.5 T MRI, three IceRod cryoncedles were placed via the transperineal approach into the left anterior prostate and freezing was performed under imaging guidance with three freeze-thaw cycles. An iceball is clearly visible on axial T2-weighted image during the freezing phase (Panel **d**, tumor arrow, iceball arrowheads, urethra dashed arrow).Subsequent post-ablation dynamic gadolinium enhanced series demonstrates the corresponding ablation zone to encompass the previously demonstrated cancerous lesion (Panel **e**, arrow)

MR Thermometry for Thermal Ablation

One major advantage of MR guidance for heating thermal ablations is ablation monitoring using MRI thermometry where subsequent dose estimations can be applied. The MR thermometry typically performed is a near real-time proton resonance frequency (PRF) sequence which demonstrates signaling change as a function of temperature [57]. During delivery of ablative energy (generated by ultrasound transducer or laser applicator a series of 2D phase sensitive T1-weighted fast spoiled gradient-recalled echo MR images are acquired on MRI scanner [58–60]. Based on temperature changes, a thermal dose can be calculated to predict a real-time tissue lethal dose [61]. However, the two major limitations with this sequence are the inaccuracy in fat tissue and susceptibility to motion artifact.

MR-Guided Laser Ablation

Laser-induced interstitial thermal therapy (LITT) uses a locally placed laser fiber probe to deliver targeted thermal ablation under MR guidance (Fig. 9.2). LITT is inherently magnetic resonance (MR) compatible. MR guidance for laser applicator placement and ablative monitoring provide the imaging to prevent encroachment onto adjacent critical structures. Two early publications demonstrated technical feasibility of laser ablation monitoring in canine prostate and demonstrated correlation of the MR temperature map with contrast enhanced T1-weighted images [62, 63]. A subsequent study in cadavers demonstrated technical feasibility in the human prostate within a 3 T MRI scanner [64]. Lee et al. examined 23 patients treated with focal laser ablation demonstrating promising results [45]. Raz et al. described using laser ablation for treatment of two prostate cancer patients at 1.5 T with discharge 3 h after the procedure [65]. These studies demonstrate the potential utility of laser

Fig. 9.2 Sixty-five-year-old male presents with Gleason 3 + 4 prostatic adenocarcinoma of the left peripheral prostate. Multiparametric MRI at 3 T demonstrated an area of decreased signal intensity in the left lateral peripheral zone at the prostatic midgland (Panel **a**, arrow) with corresponding hypointensity on ADC map (Panel **b**, arrow), and early subtle hyperenhancement (Panel **c**, arrow). InSightec ultrasound transducer was placed in the rectum and treatment plan created (Panel **d**). Post-ablation dynamic gadolinium enhanced series demonstrates the corresponding ablation zone to encompass the previously demonstrated cancerous lesion (axial Panel **e**, arrow and sagittal, Panel **f**)

ablation in the prostate. However, more clinical data is needed to determine short and long term efficacy.

US- and MR-Guided High-Intensity Focused Ultrasound (HIFU) Ablation

Treatment of the prostate with focused ultrasound ablation is not new although MRI guided version of procedure has not, as of yet, been approved by FDA in the United States. HIFU achieves cellular death by rising the cellular temperature >60 °C causing cellular necrosis. HIFU ablation technique does not require placement of a needle probe in a targeted prostate tumor via the rectum or skin (perineum) to deliver thermal energy and destroy cancerous tissue. This treatment modality has been performed with transrectal ultrasound (US) imaging guidance with success in Europe for many years [66, 67]. The major limitation of US imaging guidance for prostate ablation is that ultrasound cannot precisely visualize the focus of cancer and therefore the target of therapy. Therefore, the initial treatment strategy used with US-guided high intensity focused ultrasound (HIFU) was to ablate the entire prostate, or a relatively large region where the site of biopsy-proven cancer was found using a mapping biopsy and/or mpMRI. This often resulted in inadequate tumor control or over-ablation of unnecessary normal/neural tissue with potential subsequent morbidity. An early study, using HIFU ablation in prostate by Gelet et al., treated 82 patients who were subsequently followed up for 24-month duration. These patients also received subsequent radiation treatment [68]. Among these patients, 68% were cancer free at the time of follow-up. Due to relatively high complication rates, the treatment device underwent multiple iterations and improvements. A subsequent study, by Gelet et al., demonstrated incontinence and impotency rates around 14% and 61% respectively at 19 months post-treatment. In both studies, major limitations were identified as total procedure time due to a need to cover the entire prostate and inability to monitor temperature elevations or ablation zone expansion [69]. Current generation US guided HIFU has evolved into more robust treatment platforms with motion detection, improved planning modules and capacity to perform focal and partial gland therapy using US/MRI fusion [70]. The largest prospective single arm study with 1002 patients demonstrated that whole gland ablation could be performed with severe incontinence rates from 3% to 6% and urethral stricture rates of 6–35% [71]. The 8 year biochemical free survival rates were 76% for low risk, 63% for intermediate risk, and 57% for high risk [71]. A subsequent single arm prospective clinical trial of HIFU hemiablation in 50 patients with unilateral, low-intermediate risk disease and 39.5 month followup demonstrated biochemical recurrence rate of 28–36% with 6% incontinence rate and 20% impotence rate [72]. As the ultrasound guided technology has improved so have the results [70]. A 2 year followup of 928 patients treated with three sequential versions of Sonablate devices demonstrated a corresponding 5 year biochemical disease free rate of 48.3%, 62.3%, and 82% respectively [70]. At the current time, ultrasound guidance is challenged with the inability to see the tissue heating produced by the focused ultrasound such that there is no real-time feedback. This is one of the

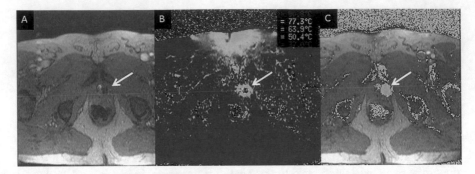

Fig. 9.3 Sixty-seven-year-old male presents with Gleason 3 + 4 prostate adenocarcinoma in left lateral peripheral zone at the apex. During MR guided laser ablation, axial T1-weighted images demonstrate a small hypointensity in left lateral peripheral zone within the lesion (Panel **a**, arrow) which corresponds to the change in temperature seen on phase imaging (Panel **b**, arrow) and calculated ablation zone on the damage map (Panel **c**, arrow)

advantages of real-time MR imaging which does in fact see and measure the tissue heating in a real-time quantitative manner.

To address the issue of visualization and temperature monitoring, two different MR guided focused ultrasound systems have been made but neither currently has FDA approval. With MR thermal monitoring and localization of lesions/zones within prostate, focused ultrasound could be performed with smaller treatment zones presumably resulting in improved treatment margins with decreased morbidity. Currently, there are two MRI-integrated systems using transrectal (Exablate, InSightec, Haifa, Israel) (Figs. 9.2 and 9.3) or transurethral (Profound Medical Inc., Toronto, Canada) transmission routes for treatment of prostate lesions with focused ultrasound technology. The system is fully integrated with the MRI console with temperature feedback control to adjust power, frequency, and rotation rate. Both systems are currently being used in patient trials assessing safety and efficacy for evidence to get FDA approval.

9.3 Recurrent Prostate Cancer

Recurrent prostate cancer after surgery can range from 25% to 40% manifesting as a rise in PSA [73–75]. Close to 30,000 men will develop biochemical recurrence (BCR) with rising PSA after radical prostatectomy each year in the USA [74]. Approximately 81% of prostate cancer recurrences occur locally in the prostate bed and can be visualized with multiparametric MRI [76]. With radiation treatment, biochemical recurrence can range widely between 33% and 63% over 10 years, and this contributes another 45,000 men/year with post-radiotherapy recurrent cancer in the USA alone [77, 78]. Salvage treatments currently available for recurrent prostate cancer include salvage radical prostatectomy, salvage radiotherapy, salvage ultrasound (U/S)-guided high-intensity focused ultrasound, salvage U/S-guided cryoablation, and newly described salvage MRI-guided laser and cryoablation.

9.3.1   MRI for Recurrent Prostate Cancer

After a definitive radical prostatectomy or radiation therapy, patients are followed at periodic intervals with measurement of PSA levels and DRE. However, DRE is frequently unreliable in evaluating local recurrent disease after radical prostatectomy. Following a radical prostatectomy, PSA levels are expected to be undetectable within several weeks of surgery. If there is a rise in a previously undetectable or stable postoperative PSA level (biochemical failure), a prompt search for persistent, recurrent, or metastatic disease should be pursued. However, PSA alone does not differentiate local from distant disease recurrence. There are three main categories of recurrence after radical prostatectomy for prostate cancer, including (1) local recurrence in the prostatic bed, (2) distant metastasis (e.g., bone, lymph node) and (3) a combination of local recurrence and distant metastasis. Therefore, the major objective of the diagnostic imaging studies is to assess patients for the presence of distant metastatic disease or local recurrent disease, each requiring different forms of systemic or local therapy. Local recurrence may be amenable to salvage therapy. Systemic recurrence may be an indication for systemic treatment including androgen deprivation therapy and/or chemotherapy.

Transrectal ultrasound (TRUS) has been used for the evaluation of local recurrence. However, the altered anatomy of the region, the development of fibrotic tissue, the fact that 30% of recurrent tumors may be isoechoic and that some lesions are in an anterior position or extend along the bladder wall influence the accuracy of this modality. Furthermore, CT imaging can depict only local recurrences of greater than or equal to 2 cc [79].

The use of biopsy has been questioned in the face of a rising PSA level, since the negative results are unreliable and elevated PSA levels usually precede clinical evidence of local recurrence by 1 or more years. Repeat TRUS with vesicourethral anastomosis (VUA) needle biopsy may be necessary to document local recurrence in one-third of cases [80]. About 25% of men with post-prostatectomy PSA levels of less than 1 ng/ml have histologic confirmation of local recurrence after biopsy of the prostatic fossa [81]. In a more contemporary study, MRI directed biopsies in 132 post-prostatectomy patients using cognitive/visual registration and TRUS guided biopsies, with a median PSA of 0.59 ng/ml and a median lesion size on MRI of 1 cm yielding a positive predictive value of 85% with positive biopsy rates of 74% with lesion sizes between 1 and 2 cm [82].

[11]C-choline PET/CT has an advantage to reveal both locally recurrent and distant metastatic malignant lesions. ^{11}C-choline PET/CT had a sensitivity of 73%, a specificity of 88%, a positive predictive value (PPV) of 92%, a negative predictive value (NPV) of 61%, and an accuracy of 78% for the detection of clinically suspected

recurrent prostate cancer in postsurgical patients [8]. However, ^{11}C-choline PET/CT is not widely available.

With the limitations of US and CT imaging, MRI has been shown to be quite useful in detection and staging of recurrent prostate tumors [83–85]. MRI provides superior soft tissue contrast resolution, high spatial resolution, multiplanar imaging capabilities, and a large field of view. The use of integrated endorectal and pelvic phased-array coils has led to improved visualization of the prostatic fossa. The use of mpMRI for recurrent prostate cancer continues to evolve and has potential to evaluate both local recurrence and distant bony and nodal metastases [9]. Functional information from MR spectroscopic imaging and diffusion-weighted imaging may complement morphologic MRI by reflecting tissue biochemistry and Brownian motion of water molecules, respectively. These functional imaging techniques may be used to supplement conventional MR imaging in diagnostic clinical studies.

9.3.2 Salvage Therapies for Prostate Cancer

9.3.2.1 Surgery

Salvage radical prostatectomy (sRP) after radiotherapy is more difficult because of local fibrosis and tissue plane changes secondary to the radiation. From this standpoint only a few centers take on these cases. However, sRP has the longest follow-up period for any of the salvage therapies with follow-up greater than 10 years. The biochemical disease free survival (bDFS) at 10 years was 30–43% based on aggregated data from four institutions. The 10 year cancer-specific survival rates were 70–77% [86, 87]. More recently, salvage robotic radical prostatectomy has been reported with some small patient studies demonstrating more promising results, but it is premature to report on long term follow up [88]. Due to the difficulties posed after primary radiation treatment failure, the complication rates for sRP have been higher than primary surgery with incontinence rates of 58% and major complication rates of 33% [89]. The largest series to date is a multi-institutional collaboration study which reviewed 404 patients with a median follow up of 4.4 years and freedom from clinical metastasis of >75% at 10 years from surgery. This study also identified the most favorable groups to undergo sRP were in men with a PSA < 4 ng/ml and post radiation prostate biopsy Gleason score of ≤7 [90].

9.3.2.2 Radiation

Salvage radiotherapy can be used for BCR following surgery or primary radiotherapy failures. Many times salvage brachytherapy (BT) is performed for primary radiotherapy failures. In a large study out of Mayo Clinic, 49 patients with primary external beam radiotherapy (EBRT) failure were treated with salvage low dose rate BT. They demonstrated a 3 year biochemical disease free survival (bDFS) of 48%

and a 5 year bDFS of 34% [91]. Multiple other studies demonstrate a slightly better bDFS, but neoadjuvant androgen treatment was also used in conjunction with the radiation confounding the results. Overall, the 5 year bDFS for salvage BT after primary radiotherapy is approximately 20–70%. Complications for salvage BT were either genitourinary (GU) or gastrointestinal (GI). Grade 3–4 GU toxicity was 17% as a late complication and grade 3–4 GI toxicity was around 5.6% [89, 91, 92]. In a more contemporary series of 98 patients, the 3 year bDFS was 60.1% and there was no difference between low dose rate BT and high dose rate BT. On multivariate analysis, only the prostate specific antigen doubling time (PSADT) <12 months was significantly associated with PSA relapse [93].

9.3.2.3   High Intensity Focused Ultrasound (HIFU)

Salvage high intensity focused ultrasound (HIFU) which targets focused ultrasound energy to a specific area has been used for primary prostate cancer treatment and for salvage therapy. Salvage HIFU is relatively recent treatment modality with limited studies on its efficacy. Three different studies have been published with relatively short follow up period of 7.4–18.1 months. These studies demonstrated a highly variable bDFS of 25–71% which was confounded by variable definitions of PSA failure and variable use of hormonal therapy before treatment. The most commonly reported complications are incontinence (10–49.5%), urethral stricture with retention (17–17.6%), erectile dysfunction (66.2–100%), and recto-urethral fistula (3–16%) [89, 94–96]. In the largest multi-institutional pooled series of 418 patients treated with whole gland HIFU after failed radiotherapy, the 7 year cancer specific survival and metastasis free survival of >80% were attained at the price of significant morbidity. According to this study, salvage HIFU should be initiated early following radiation failure and by centers with significant experience [97].

9.3.2.4   Ultrasound-Guided Cryoablation

Ultrasound guided cryotherapy is currently being used for primary prostate cancer treatment as well as salvage treatment after primary radiotherapy failure. Due to the relative recent development as a treatment modality, there are limited studies on its efficacy. Chin et al. reported on 118 patients treated with salvage US cryotherapy after radiotherapy failure [98]. This study showed a negative biopsy rate of 87% with a median follow up of 18.6 months. Siddiqui et al. presented 15 patients with salvage ultrasound guided cryotherapy after radical retropubic prostatectomy [99]. Their findings demonstrated a 40% bDFS at a mean follow up of 20 months. As cryotherapy devices have evolved with mixed gas technology, smaller cryoprobe size, improved urethral preservation with warmers, better imaging, and increased operator experience, the success rates have improved and complication rates decreased. A recent large study from the COLD cryo on-line data registry reported a 5 year bDFS to be 58.9% by the ASTRO definition of BCR and 54.5% by the

Phoenix definition of BCR [100]. For patients treated with salvage US guided cryotherapy after primary radiotherapy failure, the most recent reported complication
rates are perineal pain (4–14%), mild-moderate incontinence (6–13%), severe
incontinence (2–4%), and urethrorectal fistula (1–2%). With the use of urethral
warming catheter, the rate of sloughing and urethral stricture has been reduced to
near zero. Erectile dysfunction (ED) is still high with rates of 69–86% [89]. In a
pooled study of 396 patients who underwent salvage cryosurgery for radiation failure with a medial follow up of 47.8 months, had respective 5 and 10 year DFS of
63% and 35% with disease specific survivals of 91% and 79% respectively [101].

9.3.3   Selection of Patients for Focal Therapy

One of the most important aspects of assessing recurrent prostate cancer is determination of whether the recurrence is localized or metastatic [75]. The second issue in
managing patients with BCR of prostate cancer is assessing the risk of cancer treatment versus the risk of further intervention. Overall, rapid PSA rise, short-disease
free interval, and high-grade disease are all poor prognostic indicators with a higher
likelihood of systemic recurrence, while slow PSA rise, long disease free interval,
and low-grade disease are better prognostic indicators with a higher likelihood of
local recurrence [41, 102].

Potential criteria for MR guided focal ablative treatment of recurrent prostate
cancer are as follows: (1) biopsy proven local recurrent tumor that can be visualized
by MRI, (2) absence of distant metastasis confirmed with chest, abdomen, pelvis
CT and/or MRI plus bone scintigraphy and/or ^{11}C choline PET/CT scan [9, 42].
Although not perfect, these criteria seek to rule out patients where they have both
local and systemic metastases unless local treatment is coupled with systemic treatment strategy for cancer control.

9.4   MRI Guided Recurrent Prostate Cancer Focal Therapy Options

9.4.1   MR-Guided Cryoablation

MR guided cryoablation for recurrent prostate cancer is technically feasible and
been successful in short-term follow-up. Woodrum et al. published on 18 patients
treated with MR guided cryoablation for locally recurrent prostate cancer where
treatment optimization parameters were assessed for two groups of nine patients
[55]. Ultimately, the study demonstrated that a more aggressive tight (5 mm) spacing of cryoneedles, three freeze-thaw cycles, and prudent adjustment of the urethral
warmer temperature produced better short-term recurrence free intervals. Gangi

et al. also demonstrated successful MR-guided cryoablation treatment of several patients with recurrent prostate cancer [56]. This technique offers the advantage that it is not appreciably limited by the prior surgical or radiation treatment to the targeted area [55, 56]. Using MR guidance, cryoablation treatment can be tailored to the desired area (Figs. 9.2 and 9.4). In another series, MR-guided cryoablation has been reported to successfully treat select patients with locally recurrent tumors after failed radiation therapy [103].

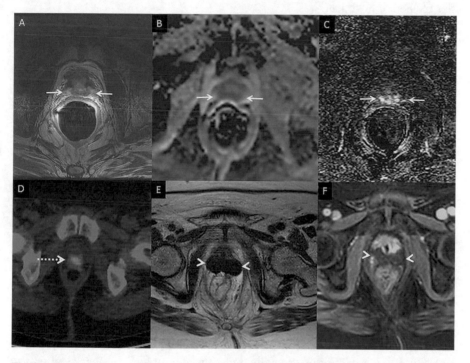

Fig. 9.4 Seventy-five-year-old male with history of refractory prostate cancer status post external beam radiation therapy with subsequent salvage prostatectomy and lymphadenectomy. PSA was undetectable but started rising. Pelvic MRI demonstrated a hyperenhancing recurrence posterior to the vesicourethral (VU) anastomosis. TRUS-guided biopsy revealed Gleason 4 + 4. Axial T2-weighted images demonstrate a soft tissue nodule with hypointensity posterior to VU anastomosis (Panel **a**, arrows). Corresponding ADC map demonstrates restricted diffusion (Panel **b**, arrows). Corresponding DCE image demonstrates hyperenhancement (Panel **c**, arrows). CT/PET choline imaging demonstrates corresponding increased activity in the hyperenhancing nodule (Panel **d**, dashed arrow). Using a 1.5 T MRI, six IceRod cryoneedles were placed via the transperineal approach posterior to the vesicourethral anastomosis. An iceball is clearly visible on axial T2-weighted image during the freezing phase (Panel **e**, iceball arrowheads).Subsequent post-ablation dynamic gadolinium enhanced series demonstrates the corresponding ablation zone to encompass the previously demonstrated cancerous lesion (Panel **f**, arrowheads)

9.4.2 MR-Guided Laser Interstitial Therapy (LITT)

Using Laser interstitial thermal therapy (LITT) for recurrent prostate cancer has been shown to be feasible with a case report using a 3 T MRI with Visualase 980 nm diode laser system (Medtronic, Minneapolis, MN, USA) [104]. A small case series was also presented by the same group which demonstrated feasibility of treating recurrent prostate cancer with laser ablation [104]. Difficulties encountered with this ablation technique in these patients were the temperature mapping distortion secondary to the surgical clips from prior surgery. This could also be encountered with brachytherapy seed implantation as well. Therefore, recurrences within the surgical clips or brachytherapy seeds would represent a relative contraindication for this method of treatment.

9.5 Follow-Up Imaging

After MR-guided salvage focal ablation, the best way to monitor the patient is by measuring serial serum PSA and MR imaging. PSA levels should decrease soon after ablation and ideally drop to undetectable within several weeks of salvage procedure if there is no remaining prostate tissue. In the setting of prior radiation, the PSA is expected to return to prior baseline PSA levels seen after radiation treatment. In either situation the PSA should decrease to a new plateau level and remain there over time. A rise in a previously undetectable or stable postoperative PSA levels during post-treatment follow up indicate recurrent or possibly metastatic disease warranting a further workup to localize viable disease.

One possible schematic for follow up is PSA every 3 months and MR imaging at 6, 12, 18, 24 months post procedure and then lengthen to yearly after the first 2 years post-ablation if all is negative. Cryoablation has been shown to have some residual ablation zone contrast enhancement when imaging less than 6 months post ablation which resolves at 6 month imaging [105]. Multiparametric MRI can assess prostatic fossa, iliac lymph nodes and pelvic bones. Mild inflammatory enhancement about the ablation zone without a discrete mass is a common finding after procedure and usually resolves within 3 months after procedure. Persistent or new discrete enhancing nodules on MRI are suspicious for residual or recurrent cancerous lesions. These enhancing nodules, if still confined in the prostatic bed, may be amenable for repeated MR-guided salvage ablation. Post ablative biopsies should also be entertained at 1 and 2 years post treatment with particular attention to the margin of the ablation zone.

9.6   Challenges of Focal Therapy

9.6.1   Limitations to MRI Visualization of Iceball Temperature Isotherms

A limitation for MR-guided cryoablation is that the leading edge of the iceball is well visualized due to very rapid T2 relaxation of ice protons, but this corresponds to 0 °C and may not be completely lethal. Therefore, it is necessary to carry the edge of the ice beyond the tumor margin by at least 5 mm assuming that iceball lethal isotherms of −40 °C are less than 5 mm from the leading edge of the iceball [106]. Complicating factors to this assumption include heat transfer from adjacent major vessels or urethral warmers [107]. Studies have shown that ultra-short echo times (UTE) can be used to visualize temperature changes within the iceball, however this technique is yet to be widely applied clinically [108–110] Confounding the need for good margin coverage is the problem of very restrictive space in and around the prostate bed with close proximity to the rectum, bladder, and external striated urethral sphincter. This small margin for error presents an ongoing challenge of balancing treatment efficacy with morbidity.

9.6.2   Limitations of MRI Thermometry

Proton resonance temperature mapping (PRF) capitalizes on the phenomenon of linear change of resonance frequency of water protons with temperature. PRF temperature mapping is a powerful tool, but it has some major limitations such as sensitivity to motion and tissue edge artifacts. PRF relies on a baseline comparison image which all subsequent images are compared. As a consequence, motion is a large problem where the baseline image alignment is disrupted causing phase registration artifacts. A method that has been proposed to alleviate this is the referenceless temperature mapping. Another potential issue is the presence of the surgical clips, which can cause metallic artifact resulting in image distortion and signal drop-out, degrading the MR images. In the native prostate, this is less of an issue, but in the post-surgical prostate bed surgical clip artifact becomes a real problem for phase change-based temperature imaging. The final major limitation with PRF-based temperature mapping is problem with tissue/fat interface. The resonance frequency is only dependent on temperature for water protons. The resonance frequency of protons in fat is different producing artifact and inaccuracy for tissue fat interfaces. Some approaches attempt to resolve this issue by the use of the so-called Dixon technique to separate MRI signals from fat and water, use PRF method on the fat-only images and use phase changes of the fat signal to correct for non-temperature-dependent phase changes [111]. This technique, however, has not, as of yet, been applied clinically.

9.7 Conclusions

Prostate cancer is the most common solid malignancy in men. As such the clinical burden is significant, prostate cancer diagnosis and treatment for new or recurrent disease will demand considerable resources and effort for years to come. mpMRI is playing a pivotal role in the diagnosis and management of this disease. MRI and ultrasound fusion for prostate biopsy guidance appear to represent the next step in timely diagnosis and navigation to clinically significant cancers. mpMRI is an effective modality in the depiction of a locally recurrent tumor after failed definitive treatment. While minimally invasive MR-guided focal ablation of native or locally recurrent prostate cancer is feasible and rapidly becoming a viable treatment alternative, there is still continued work needed to determine long term efficacy. To date, all focal therapy treatment series suffer from relatively small patient numbers with short follow-up and need for comparison to established therapies. Additionally, it is critically important that good prospective clinical trials for each treatment modality be performed to assess the advantage of each and to determine long-term efficacy.

References

1. Potosky AL, Davis WW, Hoffman RM et al (2004) Five-year outcomes after prostatectomy or radiotherapy for prostate cancer: the prostate cancer outcomes study [see comment]. J Natl Cancer Inst 96:1358–1367
2. Wallis CJD, Saskin R, Choo R et al (2016) Surgery versus radiotherapy for clinically-localized prostate cancer: a systematic review and meta-analysis. Eur Urol 70:21–30
3. Jemal A, Siegel R, Ward E et al (2006) Cancer statistics, 2006. CA Cancer J Clin 56:106–130
4. Hambrock T, Somford DM, Huisman HJ et al (2011) Relationship between apparent diffusion coefficients at 3.0-T MR imaging and Gleason grade in peripheral zone prostate cancer. Radiology 259:453–461
5. Futterer JJ, Gupta RT, Katz A et al (2014) The role of magnetic resonance imaging (MRI) in focal therapy for prostate cancer: recommendations from a consensus panel. BJU Int 113:218–227
6. Spektor M, Mathur M, Weinreb JC (2017) Standards for MRI reporting-the evolution to PI-RADS v 2.0. Transl Androl Urol 6:355–367
7. Siddiqui MM, Rais-Bahrami S, Turkbey B et al (2015) Comparison of MR/ultrasound fusion-guided biopsy with ultrasound-guided biopsy for the diagnosis of prostate cancer. JAMA 313:390–397
8. Reske SN, Blumstein NM, Glatting G (2008) [11C]choline PET/CT imaging in occult local relapse of prostate cancer after radical prostatectomy. Eur J Nucl Med Mol Imaging 35:9–17
9. Kitajima K, Murphy RC, Nathan MA et al (2014) Detection of recurrent prostate cancer after radical prostatectomy: comparison of 11C-choline PET/CT with pelvic multiparametric MR imaging with endorectal coil. J Nucl Med 55:223–232
10. Parker WP, Davis BJ, Park SS et al (2017) Identification of site-specific recurrence following primary radiation therapy for prostate cancer using C-11 choline positron emission tomography/computed tomography: a nomogram for predicting extrapelvic disease. Eur Urol 71:340–348

11. Corfield J, Perera M, Bolton D, Lawrentschuk N (2018) (68)Ga-prostate specific membrane antigen (PSMA) positron emission tomography (PET) for primary staging of high-risk prostate cancer: a systematic review. World J Urol 36:519–527
12. Afshar-Oromieh A, Holland-Letz T, Giesel FL et al (2017) Diagnostic performance of (68) Ga-PSMA-11 (HBED-CC) PET/CT in patients with recurrent prostate cancer: evaluation in 1007 patients. Eur J Nucl Med Mol Imaging 44:1258–1268
13. Wieder H, Beer AJ, Holzapfel K et al (2017) 11C-choline PET/CT and whole-body MRI including diffusion-weighted imaging for patients with recurrent prostate cancer. Oncotarget 8:66516–66527
14. Hodge KK, McNeal JE, Terris MK, Stamey TA (1989) Random systematic versus directed ultrasound guided transrectal core biopsies of the prostate. J Urol 142:71–74 discussion 4–5
15. Eichler K, Hempel S, Wilby J, Myers L, Bachmann LM, Kleijnen J (2006) Diagnostic value of systematic biopsy methods in the investigation of prostate cancer: a systematic review. J Urol 175:1605–1612
16. Jones JS (2007) Saturation biopsy for detecting and characterizing prostate cancer. BJU Int 99:1340–1344
17. Lane BR, Zippe CD, Abouassaly R, Schoenfield L, Magi-Galluzzi C, Jones JS (2008) Saturation technique does not decrease cancer detection during followup after initial prostate biopsy. J Urol 179:1746–1750 discussion 50
18. Nelson AW, Harvey RC, Parker RA, Kastner C, Doble A, Gnanapragasam VJ (2013) Repeat prostate biopsy strategies after initial negative biopsy: meta-regression comparing cancer detection of transperineal, transrectal saturation and MRI guided biopsy. PLoS One 8:e57480
19. Ahmed HU, Emberton M, Kepner G, Kepner J (2012) A biomedical engineering approach to mitigate the errors of prostate biopsy. Nat Rev Urol 9:227–231
20. Salami SS, Ben-Levi E, Yaskiv O et al (2015) In patients with a previous negative prostate biopsy and a suspicious lesion on magnetic resonance imaging, is a 12-core biopsy still necessary in addition to a targeted biopsy? BJU Int 115:562–570
21. Arumainayagam N, Ahmed HU, Moore CM et al (2013) Multiparametric MR imaging for detection of clinically significant prostate cancer: a validation cohort study with transperineal template prostate mapping as the reference standard. Radiology 268:761–769
22. Ahmed HU, Kirkham A, Arya M et al (2009) Is it time to consider a role for MRI before prostate biopsy? Nat Rev Clin Oncol 6:197–206
23. Haffner J, Lemaitre L, Puech P et al (2011) Role of magnetic resonance imaging before initial biopsy: comparison of magnetic resonance imaging-targeted and systematic biopsy for significant prostate cancer detection. BJU Int 108:E171–E178
24. Valerio M, McCartan N, Freeman A, Punwani S, Emberton M, Ahmed HU (2015) Visually directed vs. software-based targeted biopsy compared to transperineal template mapping biopsy in the detection of clinically significant prostate cancer. Urol Oncol 33:424 e9–424 16
25. Tyson MD, Arora SS, Scarpato KR, Barocas D (2016) Magnetic resonance-ultrasound fusion prostate biopsy in the diagnosis of prostate cancer. Urol Oncol 34:326–332
26. Mozer P, Roupret M, Le Cossec C et al (2015) First round of targeted biopsies using magnetic resonance imaging/ultrasonography fusion compared with conventional transrectal ultrasonography-guided biopsies for the diagnosis of localised prostate cancer. BJU Int 115:50–57
27. Meng X, Rosenkrantz AB, Mendhiratta N et al (2016) Relationship between prebiopsy multiparametric magnetic resonance imaging (MRI), biopsy indication, and MRI-ultrasound fusion-targeted prostate biopsy outcomes. Eur Urol 69:512–517
28. Sonn GA, Chang E, Natarajan S et al (2014) Value of targeted prostate biopsy using magnetic resonance-ultrasound fusion in men with prior negative biopsy and elevated prostate-specific antigen. Eur Urol 65:809–815
29. Hoeks CM, Schouten MG, Bomers JG et al (2012) Three-tesla magnetic resonance-guided prostate biopsy in men with increased prostate-specific antigen and repeated, negative, ran-

dom, systematic, transrectal ultrasound biopsies: detection of clinically significant prostate cancers. Eur Urol 62:902–909

30. Delongchamps NB, Lefevre A, Bouazza N, Beuvon F, Legman P, Cornud F (2015) Detection of significant prostate cancer with magnetic resonance targeted biopsies—should transrectal ultrasound-magnetic resonance imaging fusion guided biopsies alone be a standard of care? J Urol 193:1198–1204

31. Filson CP, Natarajan S, Margolis DJ et al (2016) Prostate cancer detection with magnetic resonance-ultrasound fusion biopsy: the role of systematic and targeted biopsies. Cancer 122:884–892

32. Borofsky S, George AK, Gaur S et al (2018) What are we missing? False-negative cancers at multiparametric MR imaging of the prostate. Radiology 286:186–195

33. Penzkofer T, Tuncali K, Fedorov A et al (2015) Transperineal in-bore 3-T MR imaging-guided prostate biopsy: a prospective clinical observational study. Radiology 274:170–180

34. Elhawary H, Zivanovic A, Rea M et al (2006) The feasibility of MR-image guided prostate biopsy using piezoceramic motors inside or near to the magnet isocentre. Med Image Comput Comput Assist Interv Int Conf 9:519–526

35. Lagerburg V, Moerland MA, van Vulpen M, Lagendijk JJW (2006) A new robotic needle insertion method to minimise attendant prostate motion. Radiother Oncol 80:73–77

36. Heidenreich A, Bastian PJ, Bellmunt J et al (2014) EAU guidelines on prostate cancer. Part 1: screening, diagnosis, and local treatment with curative intent-update 2013. Eur Urol 65:124–137

37. Lei JH, Liu LR, Wei Q et al (2015) Systematic review and meta-analysis of the survival outcomes of first-line treatment options in high-risk prostate cancer. Sci Rep 5:7713

38. Petrelli F, Vavassori I, Coinu A, Borgonovo K, Sarti E, Barni S (2014) Radical prostatectomy or radiotherapy in high-risk prostate cancer: a systematic review and metaanalysis. Clin Genitourin Cancer 12:215–224

39. Cooperberg MR, Broering JM, Carroll PR (2010) Time trends and local variation in primary treatment of localized prostate cancer. J Clin Oncol 28:1117–1123

40. Sanda MG, Cadeddu JA, Kirkby E et al (2018) Clinically localized prostate cancer: AUA/ASTRO/SUO guideline. Part II: recommended approaches and details of specific care options. J Urol pii:S0022-5347(17)78003-2

41. Hakimi AA, Feder M, Ghavamian R (2007) Minimally invasive approaches to prostate cancer: a review of the current literature. Urol J 4:130–137

42. Menon M, Tewari A, Peabody JO et al (2004) Vattikuti institute prostatectomy, a technique of robotic radical prostatectomy for management of localized carcinoma of the prostate: experience of over 1100 cases. Urol Clin North Am 31:701–717

43. Polascik TJ (2014) How to select the right patients for focal therapy of prostate cancer? Curr Opin Urol 24:203–208

44. Katz AE (2009) Prostate cryotherapy: current status. Curr Opin Urol 19:177–181

45. Lee T, Mendhiratta N, Sperling D, Lepor H (2014) Focal laser ablation for localized prostate cancer: principles, clinical trials, and our initial experience. Rev Urol 16:55–66

46. Rogenhofer S, Ganzer R, Lunz J-C, Schostak M, Wieland WF, Walter B (2008) Eight years' experience with high-intensity focused ultrasonography for treatment of localized prostate cancer. Urology 72:1329–1333 discussion 33–34

47. Cooperberg MR, D'Amico AV, Karakiewicz PI et al (2013) Management of biochemical recurrence after primary treatment of prostate cancer: a systematic review of the literature. Eur Urol 64:905–915

48. Budaus L, Spethmann J, Isbarn H et al (2011) Inverse stage migration in patients undergoing radical prostatectomy: results of 8916 European patients treated within the last decade. BJU Int 108:1256–1261

49. van den Bos W, Pinto PA, de la Rosette JJ (2014) Imaging modalities in focal therapy: patient selection, treatment guidance, and follow-up. Curr Opin Urol 24:218–224

50. Stephenson AJ, Scardino PT, Bianco FJ Jr, Eastham JA (2004) Salvage therapy for locally recurrent prostate cancer after external beam radiotherapy. Curr Treat Options in Oncol 5:357–365

51. Tacke J, Adam G, Haage P, Sellhaus B, Grosskortenhaus S, Gunther RW (2001) MR-guided percutaneous cryotherapy of the liver: in vivo evaluation with histologic correlation in an animal model. J Magn Reson Imaging 13:50–56

52. Tuncali K, Morrison PR, Tatli S, Silverman SG (2006) MRI-guided percutaneous cryoablation of renal tumors: use of external manual displacement of adjacent bowel loops. Eur J Radiol 59:198–202

53. Josan S, Bouley DM, van den Bosch M, Daniel BL, Butts PK (2009) MRI-guided cryoablation: in vivo assessment of focal canine prostate cryolesions. J Magn Reson Imaging 30:169–176

54. van den Bosch MA, Josan S, Bouley DM et al (2009) MR imaging-guided percutaneous cryoablation of the prostate in an animal model: in vivo imaging of cryoablation-induced tissue necrosis with immediate histopathologic correlation. J Vasc Interv Radiol 20:252–258

55. Woodrum DA, Kawashima A, Karnes RJ et al (2013) Magnetic resonance imaging-guided cryoablation of recurrent prostate cancer after radical prostatectomy: initial single institution experience. Urology 82:870–875

56. Gangi A, Tsoumakidou G, Abdelli O et al (2012) Percutaneous MR-guided cryoablation of prostate cancer: initial experience. Eur Radiol 22:1829–1835

57. McNichols RJ, Gowda A, Kangasniemi M, Bankson JA, Price RE, Hazle JD (2004) MR thermometry-based feedback control of laser interstitial thermal therapy at 980 nm. Lasers Surg Med 34:48–55

58. Vitkin IA, Moriarty JA, Peters RD et al (1997) Magnetic resonance imaging of temperature changes during interstitial microwave heating: a phantom study. Med Phys 24:269–277

59. Hynynen K, Freund WR, Cline HE et al (1996) A clinical, noninvasive, MR imaging-monitored ultrasound surgery method. Radiographics 16:185–195

60. Ishihara Y, Calderon A, Watanabe H, Okamoto K, Suzuki Y, Kuroda K (1995) A precise and fast temperature mapping using water proton chemical shift. Magn Reson Med 34:814–823

61. Sapareto SA, Dewey WC (1984) Thermal dose determination in cancer therapy. Int J Radiat Oncol Biol Phys 10:787–800

62. McNichols RJ, Gowda A, Gelnett MD, Stafford RJ (2005) Percutaneous MRI-guided laser thermal therapy in canine prostate. Photonic therapeutics and diagnostics. SPIE, San Jose, CA, pp 214–225

63. Stafford RJ, Shetty A, Elliott AM et al (2010) Magnetic resonance guided, focal laser induced interstitial thermal therapy in a canine prostate model. J Urol 184:1514–1520

64. Woodrum DA, Gorny KR, Mynderse LA et al (2010) Feasibility of 3.0T magnetic resonance imaging-guided laser ablation of a cadaveric prostate. Urology 75:1514.e1–1514.e6

65. Raz O, Haider MA, Davidson SR et al (2010) Real-time magnetic resonance imaging-guided focal laser therapy in patients with low-risk prostate cancer. Eur Urol 58:173–177

66. Blana A, Rogenhofer S, Ganzer R et al (2008) Eight years' experience with high-intensity focused ultrasonography for treatment of localized prostate cancer. Urology 72:1329–1333 discussion 33–34

67. Thuroff S, Chaussy C, Vallancien G et al (2003) High-intensity focused ultrasound and localized prostate cancer: efficacy results from the European multicentric study. J Endourol 17:673–677

68. Gelet A, Chapelon JY, Bouvier R et al (2000) Transrectal high-intensity focused ultrasound: minimally invasive therapy of localized prostate cancer. J Endourol 14:519–528

69. Gelet A, Chapelon JY, Bouvier R, Rouviere O, Lyonnet D, Dubernard JM (2001) Transrectal high intensity focused ultrasound for the treatment of localized prostate cancer: factors influencing the outcome. Eur Urol 40:124–129

70. Uchida T, Tomonaga T, Kim H et al (2015) Improved outcomes with advancements in high intensity focused ultrasound devices for the treatment of localized prostate cancer. J Urol 193:103–110

71. Crouzet S, Chapelon JY, Rouviere O et al (2014) Whole-gland ablation of localized prostate cancer with high-intensity focused ultrasound: oncologic outcomes and morbidity in 1002 patients. Eur Urol 65:907–914

72. van Velthoven R, Aoun F, Marcelis Q et al (2016) A prospective clinical trial of HIFU hemiablation for clinically localized prostate cancer. Prostate Cancer Prostatic Dis 19:79–83

73. Brandeis J, Pashos CL, Henning JM, Litwin MS (2000) A nationwide charge comparison of the principal treatments for early stage prostate carcinoma. Cancer 89:1792–1799

74. Moul JW (2000) Prostate specific antigen only progression of prostate cancer. J Urol 163:1632–1642

75. Stephenson AJ, Slawin KM (2004) The value of radiotherapy in treating recurrent prostate cancer after radical prostatectomy. Nat Clin Pract Urol 1:90–96

76. Sella T, Schwartz LH, Swindle PW et al (2004) Suspected local recurrence after radical prostatectomy: endorectal coil MR imaging. Radiology 231:379–385

77. Agarwal PK, Sadetsky N, Konety BR, Resnick MI, Carroll PR, Cancer of the Prostate Strategic Urological Research E (2008) Treatment failure after primary and salvage therapy for prostate cancer: likelihood, patterns of care, and outcomes. Cancer 112:307–314

78. Kuban DA, Thames HD, Levy LB et al (2003) Long-term multi-institutional analysis of stage T1-T2 prostate cancer treated with radiotherapy in the PSA era. Int J Radiat Oncol Biol Phys 57:915–928

79. Kramer S, Gorich J, Gottfried HW et al (1997) Sensitivity of computed tomography in detecting local recurrence of prostatic carcinoma following radical prostatectomy. Br J Radiol 70:995–999

80. Connolly JA, Shinohara K, Presti JC Jr, Carroll PR (1996) Local recurrence after radical prostatectomy: characteristics in size, location, and relationship to prostate-specific antigen and surgical margins. Urology 47:225–231

81. Leventis AK, Shariat SF, Slawin KM (2001) Local recurrence after radical prostatectomy: correlation of US features with prostatic fossa biopsy findings. Radiology 219:432–439

82. Linder BJ, Kawashima A, Woodrum DA et al (2014) Early localization of recurrent prostate cancer after prostatectomy by endorectal coil magnetic resonance imaging. Can J Urol 21:7283–7289

83. Roy C, Foudi F, Charton J et al (2013) Comparative sensitivities of functional MRI sequences in detection of local recurrence of prostate carcinoma after radical prostatectomy or external-beam radiotherapy. AJR Am J Roentgenol 200:W361–W368

84. Kitajima K, Hartman RP, Froemming AT, Hagen CE, Takahashi N, Kawashima A (2015) Detection of local recurrence of prostate cancer after radical prostatectomy using Endorectal coil MRI at 3 T: addition of DWI and dynamic contrast enhancement to T2-weighted MRI. AJR Am J Roentgenol 205:807–816

85. May EJ, Viers LD, Viers BR et al (2016) Prostate cancer post-treatment follow-up and recurrence evaluation. Abdom Radiol (NY) 41:862–876

86. Amling CL, Blute ML, Bergstralh EJ, Seay TM, Slezak J, Zincke H (2000) Long-term hazard of progression after radical prostatectomy for clinically localized prostate cancer: continued risk of biochemical failure after 5 years. J Urol 164:101–105

87. Bianco FJ Jr, Scardino PT, Stephenson AJ, Diblasio CJ, Fearn PA, Eastham JA (2005) Long-term oncologic results of salvage radical prostatectomy for locally recurrent prostate cancer after radiotherapy. Int J Rad Oncol Biol Phys 62:448–453

88. Boris RS, Bhandari A, Krane LS, Eun D, Kaul S, Peabody JO (2009) Salvage robotic-assisted radical prostatectomy: initial results and early report of outcomes. BJU Int 103:952–956

89. Kimura M, Mouraviev V, Tsivian M, Mayes JM, Satoh T, Polascik TJ (2010) Current salvage methods for recurrent prostate cancer after failure of primary radiotherapy. BJU Int 105:191–201

90. Qin X, Ye D (2011) Re: Daher C. Chade, Shahrokh F. Shariat, angel M. Cronin, et al. salvage radical prostatectomy for radiation-recurrent prostate cancer: a multi-institutional collaboration. Eur urol 2011;60:205–210. Eur Urol 60:e34
91. Grado GL, Collins JM, Kriegshauser JS et al (1999) Salvage brachytherapy for localized prostate cancer after radiotherapy failure. Urology 53:2–10
92. Koutrouvelis P, Hendricks F, Lailas N et al (2003) Salvage reimplantation in patient with local recurrent prostate carcinoma after brachytherapy with three dimensional computed tomography-guided permanent pararectal implant. Technol Cancer Res Treat 2:339–344
93. Kollmeier MA, McBride S, Taggar A et al (2017) Salvage brachytherapy for recurrent prostate cancer after definitive radiation therapy: a comparison of low-dose-rate and high-dose-rate brachytherapy and the importance of prostate-specific antigen doubling time. Brachytherapy 16:1091–1098
94. Zacharakis E, Ahmed HU, Ishaq A et al (2008) The feasibility and safety of high-intensity focused ultrasound as salvage therapy for recurrent prostate cancer following external beam radiotherapy. BJU Int 102:786–792
95. Murat F-J, Poissonnier L, Rabilloud M et al (2009) Mid-term results demonstrate salvage high-intensity focused ultrasound (HIFU) as an effective and acceptably morbid salvage treatment option for locally radiorecurrent prostate cancer. Eur Urol 55:640–647
96. Gelet A, Chapelon JY, Poissonnier L et al (2004) Local recurrence of prostate cancer after external beam radiotherapy: early experience of salvage therapy using high-intensity focused ultrasonography. Urology 63:625–629
97. Crouzet S, Blana A, Murat FJ et al (2017) Salvage high-intensity focused ultrasound (HIFU) for locally recurrent prostate cancer after failed radiation therapy: multi-institutional analysis of 418 patients. BJU Int 119:896–904
98. Chin JL, Pautler SE, Mouraviev V, Touma N, Moore K, Downey DB (2001) Results of salvage cryoablation of the prostate after radiation: identifying predictors of treatment failure and complications. J Urol 165:1937–1941 discussion 41–42
99. Siddiqui SA, Mynderse LA, Zincke H et al (2007) Treatment of prostate cancer local recurrence after radical retropubic prostatectomy with 17-gauge interstitial transperineal cryoablation: initial experience. Urology 70:80–85
100. Pisters LL, Rewcastle JC, Donnelly BJ, Lugnani FM, Katz AE, Jones JS (2008) Salvage prostate cryoablation: initial results from the cryo on-line data registry. J Urol 180:559–563 discussion 63–64
101. Wenske S, Quarrier S, Katz AE (2013) Salvage cryosurgery of the prostate for failure after primary radiotherapy or cryosurgery: long-term clinical, functional, and oncologic outcomes in a large cohort at a tertiary referral Centre. Eur Urol 64:1–7
102. Uchida T, Shoji S, Nakano M et al (2011) High-intensity focused ultrasound as salvage therapy for patients with recurrent prostate cancer after external beam radiation, brachytherapy or proton therapy. BJU Int 107:378–382
103. Bomers JG, Yakar D, Overduin CG et al (2013) MR imaging-guided focal cryoablation in patients with recurrent prostate cancer. Radiology 268:451–460
104. Woodrum DA, Mynderse LA, Gorny KR, Amrami KK, McNichols RJ, Callstrom MR (2011) 3.0T MR-guided laser ablation of a prostate cancer recurrence in the postsurgical prostate bed. J Vasc Interv Radiol 22:929–934
105. Porter CA, Woodrum DA, Callstrom MR et al (2010) MRI after technically successful renal cryoablation: early contrast enhancement as a common finding. Am J Roentgenol 194:790–793
106. Gage AA, Baust J (1998) Mechanisms of tissue injury in cryosurgery. Cryobiology 37:171–186
107. Favazza CP, Gorny KR, King DM et al (2014) An investigation of the effects from a urethral warming system on temperature distributions during cryoablation treatment of the prostate: a phantom study. Cryobiology 69:128–133

108. Butts K, Sinclair J, Daniel BL, Wansapura J, Pauly JM (2001) Temperature quantitation and mapping of frozen tissue. J Magn Reson Imaging 13:99–104
109. Wansapura JP, Daniel BL, Vigen KK, Butts K (2005) In vivo MR thermometry of frozen tissue using R2* and signal intensity. Acad Radiol 12:1080–1084
110. Lu A, Daniel BL, Pauly JM, Pauly KB (2008) Improved slice selection for R2* mapping during cryoablation with eddy current compensation. J Magn Reson Imaging 28:190–198
111. Soher BJ, Wyatt C, Reeder SB, MacFall JR (2010) Noninvasive temperature mapping with MRI using chemical shift water-fat separation. Magn Reson Med 63:1238–1246

Chapter 10
Immunodiagnostics and Immunotherapy Possibilities for Prostate Cancer

Heide Schatten

Abstract Despite significant progress in early detection and improved treatment modalities prostate cancer remains the second leading cause of cancer death in American men which results in about 30,000 deaths per year in the USA. An aggressive phenotype leading to 2.58% risk of dying from prostate cancer still exists and immunotherapy has offered new possibilities to treat metastatic prostate cancer that cannot be treated by other modalities. Cancer immunotherapy is a rapidly growing field of research aimed at identifying biomarkers in immunodiagnosis and to develop new therapies by enabling the immune system to detect and destroy cancer cells. Immunotherapy falls into three different broad categories which are checkpoint inhibitors, cytokines, and vaccine immunotherapy. While immunotherapy to treat prostate cancer is still limited progress has been made; for treatment of advanced prostate cancer sipuleucel-T has been administered to patients in personalized doses to destroy prostate cancer cells which is promising and invites further research to determine immunotherapies for advanced prostate cancer. Antibody-based targeted immunotherapy and dendritic-cell-based vaccination are among the therapies that are currently being evaluated as promising approaches to treat prostate cancer. Combination immunotherapies include prostate cancer vaccines and radiotherapy for castration resistant prostate cancer. Microbial vectors for prostate cancer immunotherapy have been developed and bacterial strains have been engineered to express cancer-specific antigens, cytokines, and prodrug-converting cytokines. These approaches are addressed in the present review.

Keywords Prostate cancer · Diagnosis · Immunotherapy · Metastasis

H. Schatten (✉)
Department of Veterinary Pathobiology, University of Missouri, Columbia, MO, USA
e-mail: SchattenH@missouri.edu

© Springer Nature Switzerland AG 2018
H. Schatten (ed.), *Molecular & Diagnostic Imaging in Prostate Cancer*,
Advances in Experimental Medicine and Biology 1096,
https://doi.org/10.1007/978-3-319-99286-0_10

185

10.1 Introduction to Immunodiagnostics and Immunotherapy

Cancer immunotherapy is a rapidly growing field of research aimed at identifying biomarkers in immunodiagnosis and to develop new therapies by enabling the immune system to detect and destroy cancer cells. So far, several successful clinical applications have been reported for a subset of patients treated for various cancer types and others are currently being developed, as will be discussed below.

Immunotherapy uses various approaches to manipulate a patient's immune system to effectively detect and destroy cancer cells that are normally not recognized by the immune system. Three main approaches have been pursued in recent years which fall in three different broad categories:

1. **Checkpoint inhibitors** are used to disrupt signals that allow cancer cells to be protected from immune attacks, taking advantage of well-known information that certain protein receptors on the surface of T-cells are able to distinguish healthy cells from cancer cells. Of these, PD-1 and CTLA-4 receptors have been best studied and current checkpoint inhibitor drugs target the PD-1 and CTLA-4 receptors. Checkpoint inhibitor therapy will be discussed in more detail below.
2. **Cytokines** are used to help regulate and direct the immune system. Cytokines can be utilized for immunotherapy by injecting synthesized cytokines into a patient's body using either (a) IL-2 (Interleukin 2) designed to target the adaptive immune system (T cells, B cells) to produce more antibodies, or (b) IFN-alpha (Interferon-alpha) to help the patient's body generate innate immune cells such as dendritic cells and macrophages to attack cancer cells. Potential side effects can be encountered with cytokine therapy including depression, flu-like symptoms, and fatigue. Cytokine immunotherapy has been approved to treat circulatory cancers such as leukemia and lymphoma, as well as melanoma, bladder, and kidney cancers.
3. **Vaccine immunotherapy** is applied to build up antibodies to protect against specific cancers including cervical, prostate and bladder cancer. This approach has been especially successful to protect against cervical cancer, which resulted in HPV vaccines that have been approved by the FDA. Gardasil and Cevarix prevent infection with two high-risk HPV strains that cause 70% of cervical cancer. The mechanisms underlying development of cervical cancer have been well investigated ([1, 2]; reviewed in [3, 4]); these studies showed that in cervical cancer, infection with 'high-risk' human papillomavirus (HPV) types, such as HPV16 and HPV18, is associated with more than 90% of cervical cancer cases. The E6 and E7 oncoproteins of HPV16 induce mitotic defects by uncoupling centrosome duplication from the cell cycle while the E6 and E7 proteins of low-risk HPV6 do not induce chromosomal abnormalities and are not typically associated with malignancy ([1, 2]; reviewed in [3, 4]). Similar detailed studies so far are not available for prostate cancer and it is not yet known whether or not a virus component may play a role in prostate cancer.

As mentioned above, much research has been devoted to checkpoint immuno-therapies that are the most rapidly advancing areas of cancer immunology. The goal of immunotherapy is to increase the strength of the immune responses against tumors which can either be achieved by stimulating the activities of specific compo-nents of the immune system or counteract signals produced by cancer cells that suppress immune responses.

Tumor-infiltrating immune cells have been identified in a variety of cancers including lung [5], colorectal [6], breast [7] and head-and-neck squamous cell car-cinoma (SCCHN) [8, 9]. Markers have been determined for all major immune cells which are B cells, T cells, cytotoxic T cells, NK cells, macrophages, dendritic cells.

For immunotherapy, two immunosuppressive pathways have been identified as effective targets, the pathway of cytotoxic T lymphocyte antigen 4 (CTLA4), a mol-ecule expressed by T cells inhibiting T cell function, and the programmed cell death 1 ligand 1 (PD-L1 or B7H1) pathway used by tumor cells to inhibit the anti-tumoral immune response.

Checkpoint blockade immunotherapy has offered significant new potential for treatment of many cancer types and has been applied to several cancers including treatment of cutaneous melanoma [10], ovarian cancer [11], breast cancer [12] Merkel cell carcinoma [13], and others.

A significant body of research has been devoted to immunotherapies against melanoma but new immunotherapy treatment options for other cancers are being explored in various laboratories. Challenges remain and include that treatment is patient specific and only a limited subset of patients can benefit from new immuno-therapies. Certain biomarkers predict the benefit for individual patients and include the presence of CD8$^+$ T cells within the tumor microenvironment or tumor margins and up-regulation of PD-1. Biomarkers are used to help determine which patients are more likely to respond to checkpoint inhibitor therapy PD-L1 and a genetic feature called microsatellite instability to which each person can respond differently.

PD-1 receptor seeks out PD-L1 protein on cells to determine if they are healthy. Cancer cells disguise themselves by sending a signal to PD-1 receptor to disrupt the signal to PD-1 using the protein PD1. Several drugs have been developed such as pembrolizumab and nivolumab, a tezolizumab targeting PD-1 receptor, disrupting the signal from PD-L1, exposing the cancer cells for attack. CTLA-4 receptor helps the immune system target cancer cells; CTLA-4 receptor gets rewired by the drug ipilimumab to stimulate immune attacks.

To date, FDA approved drugs used as checkpoint inhibitors include ipilimumab, an anti-CTLA-4 antibody and nivolumab and pembrolizumab, antibodies against PD-1 (programmed death-1) that can be used either alone or in combination [14–16]. Immunotherapy treatments are durable and can increase long-survival times. About 30% of patients can benefit from PD-1 treatment spanning cancers such as melanoma, lung, head and neck, and renal cancers. Combibation immunotherapy using CTLA-4 therapy with PD-1 checkpoint blockage resulted in increased treat-ment success of malignant melanoma to 57%. The variation in treatment results in different patients is under active investigation and much research is focused on the absence or presence of tumor-specific T cells. The presence of CD8+ cells within the tumor or tumor margins have been correlated with response to PD-1 inhibition [17] as proposed in earlier assessments of immunotherapy [18–20]. The presence of CD8+ T cells has also been linked to a type I interferon (IFN-α/β) presence [21, 22]. These observations resulted in classification of patients into T-cell-inflamed (positive for CD8+ T-cell and type I interferon presence) and non-T-cell inflamed (lacking CD8+ T-cell and type I interferon presence) (reviewed in [23]).

Escape mechanisms either in the T-cell inflamed or the non-T-cell-inflamed TME have been discussed in detail by Spranger [23] and include immune suppression and escape against an endogenous anti-tumor immune response. Side effects resulting from checkpoint inhibitors have been discussed by Wang et al. [24] and include diarrhea and colitis in patients with advanced malignancies. Other side effects can occur with immunotherapy which includes cytokine release syndrome (CRS) and neurological problems but these can be managed. Other side effects include fatigue, nausea, mouth sores, diarrhea, high blood pressure, fluid buildup but these side effects become less severe after the first treatment.

Other challenges include resistance resulting from immunotherapy; immunotherapeutic interventions are being studied in order to develop strategies to avoid or reverse resistance resulting from immunotherapy treatment.

Among the highly successful immunotherapy approaches is CART-cell (chimeric antigen receptor (CAR) T-cell) therapy which is an adaptive therapy that allows genetic reprogramming of a patient's own immune cells to detect and destroy cancer cells throughout the body. CART-cell therapy has been utilized successfully for childhood acute lymphoblastic leukemia (ALL) and for the large B-cell lymphoma (DLBCL), a subtype of non-Hodgkin lymphoma, as well as multiple myeloma.

10.2   Immunotherapy Possibilities for Prostate Cancer

Immunotherapy to treat prostate cancer is still limited but progress has been made in immunodiagnosis and immunotherapy. For treatment of advanced prostate cancer sipuleucel-T has been administered to patients in personalized doses using the patient's own immune cells that have been modified to destroy prostate cancer cells. These approaches are promising and invite further research to determine immunotherapies for advanced prostate cancer.

Despite significant progress in early detection and improved treatment modalities (reviewed in [25]) prostate cancer remains the second leading cause of cancer death in American men [26] which results in about 30,000 deaths per year in the USA [27]. An aggressive phenotype leading to 2.58% risk of dying from prostate cancer still exists [26, 28] and immunotherapy has offered new possibilities to treat metastatic prostate cancer that cannot be treated by other modalities.

While several treatment options are available for prostate cancer (reviewed in [25]) it remains a significant health problem for men in the Western world and new treatment approaches are still urgently needed. Immunotherapy presents a valuable alternative to other treatments by activating immune response to detect and destroy prostate cancer cells which so far has been possible with FDA-approved sipuleucel-T, an immunotherapeutic agent for treatment of patients with asymptomatic or minimally symptomatic castration-resistant prostate cancer (CRPC) (reviewed by [29]). This treatment and others that are currently being developed have opened new strategies with long-lasting effects to combat prostate cancer. As discussed in Chap. 3 of the companion book on prostate cancer (Cell and Molecular Biology of Prostate Cancer: Updates, Insights and New Frontiers) chronic inflammation has been linked to development of prostate cancer; inflammation of the prostate is primarily regulated by T cells, B cells, and macrophages [30–32] which functions as infiltrating immune cells during inflammation and could potentially be utilized for immunotherapy. However, it is not clear why precancerous cells can escape from the immune surveillance; several hypotheses have been proposed as discussed by Karan et al. [29] including weak immunogenicity, lack of proper communication between immune cells and precancerous cells, reduced expression of major histocompatibility complex molecules, immunologic tolerance, increased levels of regulatory T cells and activities by neoplastic cells that suppress immune activities [29]. The evasion of immune surveillance may be among the factors leading to the development of cancer.

Several immunotherapy approaches have been pursued to stimulate the immune response in the tumor microenvironment [33–37] and several antigens have been identified for immune-based therapies (reviewed in [29]). Clinical trials are in progress to pursue these new possibilities for treatment of prostate cancer [38–41]. Such new approaches include antibody-based targeted immunotherapy and dendritic-cell-based vaccination.

10.2.1   Antibody-Based Targeted Immunotherapy

For prostate cancer, CTL-associated antigen 4 (CTLA4) immunotherapy is possible [42–45]. The CTLA4 molecule is localized to the surface of T cells with functions in suppressing T cell activation. As circulating T cells display increased expression of CTLA4 in cancer patients inhibition of CTLA4 is the goal for effective immunotherapy. Two antibodies to CTLA4, ipilimumab and tremelimumab, have been applied for treatment of advanced melanoma with ipilimumab showing overall improved survival compared to standard therapy which had encouraged use of ipilimumab to treat patients with castration resistant prostate cancer (CRPC) (reviewed in [29]). Combination therapies with either PROSTVAC or GVAX prostate cancer vaccines or radiotherapy are under evaluation and proposed as promising therapies for CRPC.

10.2.2 Dendritic-Cell-Based Vaccination

Dendritic cells play a central role in the development of T effector cells which encouraged development of active immunization methods to include TAA-specific CTLs using ex vivo dendritic cells loaded with antigens or viral vaccines. The results are promising and demonstrated an induction of antigen-specific immune responses [46]. One study used autologous dendritic cells from patients with CRPC that were loaded with antigen peptides from different PSAs which resulted in increased CTL responses against PSAs.

Other studies have been performed using sipuleucel-T, an immunotherapy approved by the FDA for treatment of asymptomatic or minimally symptomatic CRPC, designed to stimulate a patient's own immune response against prostate cancer. It is a dendritic-cell-based vaccine immunotherapy which had advanced to phase II studies to give a median survival advantage of 4.1 months, with a 3-year survival rate of 31.7%, compared to 23.0% for placebo-receiving patients [47]. However, it is not clear how the treatment may benefit some patients while not others, given the observation that many of the immunotherapy treatments are patient-specific, as discussed above.

Taken together, while promising approaches are being proposed, more research and clinical tests are needed for clear answers on specific benefits for specific patients with metastatic prostate cancer.

Microbial vectors for prostate cancer immunotherapy have been developed and bacterial strains have been engineered to express cancer-specific antigens, cytokines, and prodrug-converting cytokines (reviewed in [48–59]). Bacterial strains that have been and are being developed include Listeria monocytogenes, Salmonella, Escherichia coli, and Shigella that have been genetically modified to be non-toxic to patients. These attenuated bacteria have been used successfully in preclinical animal models (reviewed in [48–60]). Vaccine approaches utilizing microbial vectors include PROSTVAC®-VF and adenovirus-based cancer vaccines to induce active immunization in murine models.

Multigene-targeted immunotherapy has been used to expand the range of anti-tumor responses by, for example, loading dendritic cells with multiple peptides. Another approach has been to load dendritic cells with a virus-expressed cDNA library specific to prostate cancer using a preclinical mouse model to gain therapeutic benefits for prostate cancer. These are a few of the many different approaches using microbial vectors for immunotherapy of prostate cancer (reviewed in [29]).

As mentioned above, combining immunotherapy with other treatment modalities are actively being pursued for optimal benefits. It includes combining immunotherapy with an adjuvant chemotherapeutic drug, radiation therapy, and immuno-suppressants.

Taken together, multiple approaches for immunotherapy have emerged and are being pursued actively. Successful applications have raised realistic expectations for continued efforts to trick the immune system into detecting and destroying tumor cells including prostate tumor cells and tissue.

10.3   Conclusion

Despite significant progress in early detection and improved treatment modalities prostate cancer remains the second leading cause of cancer death in American men which results in about 30,000 deaths per year in the USA. An aggressive phenotype leading to 2.58% risk of dying from prostate cancer still exists and immunotherapy has offered new possibilities to treat metastatic prostate cancer that cannot be treated by other modalities (reviewed in [61]).

Cancer immunotherapy is a rapidly growing field of research aimed at identifying biomarkers in immunodiagnosis and to develop new therapies by enabling the immune system to detect and destroy cancer cells. Immunotherapy falls into three different broad categories which are checkpoint inhibitors, cytokines, and vaccine immunotherapy. While immunotherapy to treat prostate cancer is still limited progress has been made; for treatment of advanced prostate cancer sipuleucel-T has been administered to patients in personalized doses to destroy prostate cancer cells which is promising and invites further research to determine immunotherapies for advanced prostate cancer. Antibody-based targeted immunotherapy and dendritic-cell-based vaccination are among the therapies that are currently being evaluated as promising approaches to treat prostate cancer. Combination immunotherapies include prostate cancer vaccines and radiotherapy for castration resistant prostate cancer. Microbial vectors for prostate cancer immunotherapy have been developed and bacterial strains have been engineered to express cancer-specific antigens, cytokines, and prodrug-converting cytokines. These approaches are promising to further develop into effective therapies to treat advanced stages of prostate cancer that currently cannot be cured by available treatment modalities.

References

1. Duensing S, Munger K (2003) Centrosome abnormalities and genomic instability induced by human papillomavirus oncoproteins. Prog Cell Cycle Res 5:383–391
2. Duensing S, Lee LY, Duensing A, Basile J, Piboonniyom S, Gonzalez S, Crum CP, Munger K (2000) The human papillomavirus type 16 E6 and E7 oncoproteins cooperate to induce mitotic defects and genomic instability by uncoupling centrosome duplication from the cell division cycle. Proc Natl Acad Sci U S A 97:10002–10007
3. Korzeniewski N, Duensing S (2012) Disruption of centrosome duplication control and induction of mitotic instability by the high-risk human papillomavirus oncoproteins E6 and E7. In: Schatten H (ed) The centrosome, Chap 12. Springer, New York, NY
4. Schatten H (2013) The impact of centrosome abnormalities on breast cancer development and progression with a focus on targeting centrosomes for breast cancer therapy. In: Schatten H (ed) Cell and molecular biology of breast cancer. Springer, New York, NY
5. Kadota K, Nitadori J, Ujiie H, Buitrago DH, Woo KM, Sima CS, Travis WD, Jones DR, Adusumilli PS (2015) Prognostic impact of immune microenvironment in lung squamous cell carcinoma: tumor-infiltrating CD10+ neutrophil/CD20+ lymphocyte ratio as an independent prognostic factor. J Thorac Oncol 10:1301–1310. https://doi.org/10.1097/JTO.0000000000000617

6. Dahlin AM, Henriksson ML, Van Guelpen B, Stenling R, Oberg A, Rutegård J, Palmqvist R (2011) Colorectal cancer prognosis depends on T-cell infiltration and molecular characteristics of the tumor. Mod Pathol 24:671–682. https://doi.org/10.1038/modpathol.2010.234

7. Adams S, Gray RJ, Demaria S, Goldstein L, Perez EA, Shulman LN, Martino S, Wang M, Jones VE, Saphner TJ, Wolff AC, Wood WC, Davidson NE et al (2014) Prognostic value of tumor-infiltrating lymphocytes in triple-negative breast cancers from two phase III randomized adjuvant breast cancer trials: ECOG 2197 and ECOG 1199. J Clin Oncol 32:2959–2966. https://doi.org/10.1200/JCO.2013.55.0491

8. Karpathiou G, Monaya A, Forest F, Froudarakis M, Casteillo F, Dumollard JM, Prades JM, Peoc'h M (2016) P16 and p53 expression status in head and neck squamous cell carcinoma: a correlation with histologic, histoprognostic and clinical parameters. Pathology 48:341–348. https://doi.org/10.1016/j.pathol.2016.01.005

9. Karpathiou G, Giroult J, Forest F, Fournel P, Monaya A, Froudarakis M, Dumollard J, Prades J, Gavid M, Peoc'h M (2016) Clinical and histological predictive factors of response to induction chemotherapy in head and neck squamous cell carcinoma. Am J Clin Pathol. https://doi.org/10.1093/ajcp/aqw145

10. Sullivan RJ, Atkins MB, Kirkwood JM, Agarwala SS, Clark JI, Ernstoff MS, Fecher L, Gajewski TF, Gastman B, Lawson DH, Lutzky J, DF MD, Margolin KA, Mehnert JM, Pavlick AC, Richards JM, Rubin KM, Sharfman W, Silverstein S, Slingluff CL Jr, Sondak VK, Tarhini AA, Thompson JA, Urba WJ, White RL, Whitman ED, Hodi FS, Kaufman HL (2018) An update on the Society for Immunotherapy of Cancer consensus statement on tumor immunotherapy for the treatment of cutaneous melanoma: version 2.0. J ImmunoTher Cancer 6:44. https://doi.org/10.1186/s40425-018-0362-6

11. Eggink LL, Roby KF, Cote R, Hoober JK (2018) An innovative immunotherapeutic strategy for ovarian cancer: CLEC10A and glycomimetic peptides. J ImmunoTher Cancer 6:28. https://doi.org/10.1186/s40425-018-0339-5

12. Bajgain P, Tawinwung S, D'Elia L, Sukumaran S, Watanabe N, Hoyos V, Lulla P, Brenner MK, Leen AM, Vera JF (2018) CAR T cell therapy for breast cancer: harnessing the tumor milieu to drive T cell activation. J ImmunoTher Cancer 6:34. https://doi.org/10.1186/s40425-018-0347-5

13. Chan IS, Bhatia S, Kaufman HL, Lipson EJ (2018) Immunotherapy for Merkel cell carcinoma: a turning point in patient care. J ImmunoTher Cancer 6:23. https://doi.org/10.1186/s40425-018-0335-9

14. Hodi FS, O'Day SJ, McDermott DF et al (2010) Improved survival with ipilimumab in patients with metastatic melanoma. N Engl J Med 363:711

15. Topalian SL, Sznol M, McDermott DF et al (2014) Survival, durable tumor remission, and long-term safety in patients with advanced melanoma receiving nivolumab. J Clin Oncol 32:1020

16. Topalian SL, Hodi FS, Brahmer JR et al (2012) Safety, activity, and immune correlates of anti-PD-1 antibody in cancer. N Engl J Med 366:2443

17. Tumeh PC, Harview CL, Yearley JH et al (2014) PD-1 blockade induces responses by inhibiting adaptive immune resistance. Nature 515:568

18. Galon J, Costes A, Sanchez-Cabo F et al (2006) Type, density, and location of immune cells within human colorectal tumors predict clinical outcome. Science 313:1960

19. Naito Y, Saito K, Shiiba K et al (1998) CD8+ T cells infiltrated within cancer cell nests as a prognostic factor in human colorectal cancer. Cancer Res 58:3491

20. Ropponen KM, Eskelinen MJ, Lipponen PK, Alhava E, Kosma VM (1997) Prognostic value of tumour-infiltrating lymphocytes (TILs) in colorectal cancer. J Pathol 182:318

21. Harlin H, Meng Y, Peterson AC et al (2009) Chemokine expression in melanoma metastases associated with CD8+ T-cell recruitment. Cancer Res 69:3077

22. Salerno EP, Olson WC, McSkimming C, Shea S, Slingluff CL Jr (2014) T cells in the human metastatic melanoma microenvironment express site-specific homing receptors and retention integrins. Int J Cancer 134:563

23. Spranger S (2016) Mechanisms of tumor escape in the context of the T-cell-inflamed and the non-T-cell-inflamed tumor microenvironment. Int Immunol 28(8):383–391. https://doi.org/10.1093/intimm/dxw014
24. Wang Y, Abu-Sbeih H, Mao E, Ali N, Ali FS, Qiao W, Lum P, Raju G, Shuttlesworth G, Stroehlein J, Diab A (2018) Immune-checkpoint inhibitor-induced diarrhea and colitis in patients with advanced malignancies: retrospective review at MD Anderson. J ImmunoTher Cancer 6:37. https://doi.org/10.1186/s40425-018-0346-6
25. Schatten H (2018) Brief overview of prostate cancer statistics, grading, diagnosis and treatment strategies. In: Schatten H (ed) Cell and molecular biology of prostate cancer: updates, insights and new frontiers. Springer, New York, NY
26. Howlader N, Noone AM, Krapcho M (eds), et al. (2015). SEER cancer statistics review, 1975–2012, National Cancer Institute: Bethesda, MD http://seer.cancer.gov/csr/1975_2012/. Accessed 5 May 2015.
27. Siegel R, Naishadham D, Jemal A (2012) Cancer statistics, 2012. CA Cancer J Clin 62:10–29
28. U.S. Cancer Statistics Working Group (2014) United States cancer statistics: 1999–2011 incidence and mortality web-based report. U.S. Department of Health and Human Services, Centers for Disease Control and Prevention and National Cancer Institute, Atlanta Available from: www.cdc.gov/uscs. Accessed 5 May 2015
29. Karan D, Holzbeierlein JM, Van Veldhuizen P, Thrasher JB (2012) Cancer immunotherapy: a paradigm shift for prostate cancer treatment. Nat Rev Urol 9:376–385
30. Robert G et al (2009) Inflammation in benign prostatic hyperplasia: a 282 patients' immuno-histochemical analysis. Prostate 69:1774–1780
31. Sottnik JL, Zhang J, Macoska JA, Keller ET (2011) The PCa tumor microenvironment. Cancer Microenviron 4:283–297
32. Theyer G et al (1992) Phenotypic characterization of infiltrating leukocytes in benign prostatic hyperplasia. Lab Invest 66:96–107
33. Dalgleish AG, Whelan MA (2006) Cancer vaccines as a therapeutic modality: the long trek. Cancer Immunol Immunother 55:1025–1032
34. Dupont B (2002) Introduction: current concepts in immunity to human cancer and therapeutic antitumor vaccines. Immunol Rev 188:5–8
35. Karan D, Krieg AM, Lubaroff DM (2007) Paradoxical enhancement of CD8 T cell-dependent anti-tumor protection despite reduced CD8 T cell responses with addition of a TLR9 agonist to a tumor vaccine. Int J Cancer 121:1520–1528
36. Saenz-Badillos J, Amin SP, Granstein RD (2001) RNA as a tumor vaccine: a review of the literature. Exp Dermatol 10:143–154
37. Schirrmacher V (1995) Tumor vaccine design: concepts, mechanisms, and efficacy testing. Int Arch Allergy Immunol 108:340–344
38. Fong L, Small EJ (2007) Immunotherapy for prostate cancer. Curr Oncol Rep 9:226–233
39. Minev BR, Guo F, Gueorguieva I, Kaiser HE (2002) Vaccines for immunotherapy of breast cancer and prostate cancer: new developments and comparative aspects. In Vivo 16:405–415
40. Simons JW, Sacks N (2006) Granulocyte-macrophage colony-stimulating factor-transduced allogeneic cancer cellular immunotherapy: the GVAX vaccine for prostate cancer. Urol Oncol 24:419–424
41. US National Library of Medicine (2012) ClinicalTrials.gov [online]. Randomized phase II trial of a DNA vaccine encoding prostatic acid phosphatase (pTVG-HP) versus GM-CSF adjuvant in patients with non-metastatic prostate cancer. US National Library of Medicine, Bethesda, MD
42. Chambers CA, Kuhns MS, Egen JG, Allison JP (2001) CTLA-4-mediated inhibition in regulation of T cell responses: mechanisms and manipulation in tumor immunotherapy. Annu Rev Immunol 19:565–594
43. May KF Jr, Gulley JL, Drake CG, Dranoff G, Kantoff PW (2011) Prostate cancer immunotherapy. Clin Cancer Res 17:5233–5238

44. Sharma P, Wagner K, Wolchok JD, Allison JP (2011) Novel cancer immunotherapy agents with survival benefit: recent successes and next steps. Nat Rev Cancer 11:805–812
45. Small EJ et al (2007) A pilot trial of CTLA-4 blockade with human anti-CTLA-4 in patients with hormone-refractory prostate cancer. Clin Cancer Res 13:1810–1815
46. Thomas-Kaskel AK, Waller CF, Schultze-Seemann W, Veelken H (2007) Immunotherapy with dendritic cells for prostate cancer. Int J Cancer 121:467–473
47. Kantoff PW et al (2010) Sipuleucel-T immunotherapy for castration-resistant prostate cancer. N Engl J Med 363:411–422
48. Bermudes D, Low KB, Pawelek J, Feng M, Belcourt M, Zheng LM, King I (2001) Tumour-selective Salmonella based cancer therapy. Biotechnol Genet Eng Rev 18:219–233
49. Bermudes D, Zheng LM, King IC (2002) Live bacteria as anticancer agents and tumor-selective protein delivery vectors. Curr Opin Drug Discov Devel 5:194–199
50. Eisenstark A, Kazmierczak RA, Fea A, Khreis R, Newman D, Schatten H (2007) Development of Salmonella strains as cancer therapy agents and testing in tumor cell lines. In: Schatten H, Eisenstark A (eds) Methods in molecular biology, Salmonella protocols, vol 253. Humana, Totowa, NJ, pp 321–353
51. Forbes NS (2010) Engineering the perfect (bacterial) cancer therapy. Nat Rev Cancer 10(11):785–794
52. Forbes NS, Munn LL, Fukumura D, Jain RK (2003) Sparse initial entrapment of systemically injected Salmonella typhimurium leads to heterogeneous accumulation within tumors. Cancer Res 63:5188–5193
53. Kazmierczak RA, Dino A, Eisenstark A, Schatten H (2013) New breast cancer treatment considerations – a brief review of the use of genetically modified (attenuated) bacteria as therapy for advanced and metastatic breast cancer. In: Schatten H (ed) Cell and molecular biology of breast cancer. Springer, New York, NY
54. Kazmierczak RA, Gentry B, Mumm T, Schatten H, Eisenstark A (2016) Salmonella bacterial monotherapy reduces autochthonous prostate tumor burden in the TRAMP mouse model. PLoS One 11(8):e0160926. https://doi.org/10.1371/journal.pone.0160926 PMID: 27504973 Free PMC Article
55. Low KB, Ittensohn M, Le T, Platt J, Sodi S, Amoss M, Ash O, Carmichael E, Chakraborty A, Fischer J, Lin SL, Luo X, Miller SI, Zheng L, King I, Pawelek JM, Bermudes D (1999) Lipid A mutant Salmonella with suppressed virulence and TNFα induction retain tumor targeting in vivo. Nat Biotechnol 17:37–41
56. Pawelek JM, Low KB, Bermudes D (1997) Tumor targeted Salmonella as a novel anticancer vector. Cancer Res 57:4537–4544
57. Pawelek JM, Sodi S, Chakraborty AK, Platt JT, Miller S, Holden DW, Hensel M, Low KB (2002) Salmonella pathogenicity island-2 and anticancer activity in mice. Cancer Gene Ther 9:813–818
58. Pawelek JM, Low KB, Bermudes D (2003) Bacteria as tumour-targeting vectors. Lancet Oncol 4:548–556
59. Saltzman DA (2005) Cancer immunotherapy based on the killing of Salmonella typhimurium-infected tumour cells. Expert Opin Biol Ther 5:443–449
60. Paterson Y, Guirnalda PD, Wood LM (2010) Listeria and Salmonella bacterial vectors of tumor-associated antigens for cancer immunotherapy. Semin Immunol 22:183–189
61. Schatten H, Ripple M (2018) The impact of centrosome pathologies on prostate cancer development and progression. In: Schatten H (ed) Cell and molecular biology of prostate cancer: updates, insights and new frontiers. Springer, New York, NY

Printed in the United States
By Bookmasters